AF528046

Silkworm Rearing

Silkworm Rearing

S.K. Ananthanarayanan

BIOTECH

2025

Biotech Books

© 2025 Reserved

ISBN10 81-7622-191-0
ISBN13 978-81-7622-191-7
EAN: 9788176221917

Information contained in this work has been published by Biotech Books and has been obtained by its author(s)/editor(s) from sources believed to be reliable and are correct to the best of their knowledge. However, the publisher and its author(s) make no representation of warranties with respect of accuracy or completeness of the contents of this book, and shall in no event be liable for any errors, omissions or damages arising out of use of this information and specifically disclaim any implied warranties or merchantability or fitness for any particular purpose.

All rights reserved. Including the right to translate or to reproduce this book or parts thereof except for brief quotations in critical reviews.

Published by : **BIOTECH BOOKS**
4762-63/23, Ansari Road, Darya Ganj,
New Delhi - 110 002
Phone : 011-4300 3222
Email : biotechbooks@yahoo.co.in

PRINTED IN INDIA

Preface

Sericulture today is one of the most prolific and important textile cultures, and better technology and awareness has only led to the advent of a book in the silk industry. Rearing silkworms is the first basic step to the sericultural process, and it is with the intention of providing a manual to farmers on rearing silkworms that this book has been written.

The book elaborates on the whole process and the steps of cultivating silkworms for producing good quality silk. It describes the morphology, anatomy and the embryology of the *Bombyx mori* silkworm, the most widely used variety of species, and explains all the procedures from the production of silkworm egg to the reeling of silk from the cocoon. It details the necessary climatic conditions, testing of eggs and their incubations, the environment in which these silkworms are to be brought up in, mulberry leaf diets, hygiene, and other necessary provisions.

Being a comprehensive compendium on silkworms rearing, this book should prove to be highly useful for farmers and agriculturists involved in sericulture.

Editor

Contents

1

Morphology and Anatomy of Silkworm

Silk is a fibrous protein of animal origin. Nearly 95% of the commercial silk in India comes from the mulberry silkworm, *Bombyx mori*.

Types of Silkworms

Silkworms belong to two families viz., Bombycidae and Saturniidae (Bombycoidea).

Bombycidae	Saturniidae
Wings are without transparent	Wings are with circular or crescent shaped transparent membrane.
Cocoon is small and yield continuous filament	Cocoon is long and bag like. Gives both continuous (tasar) and discontinuous (eri) filaments
Eg. Mulberry silkworm *B. mori*	Eg. Non-mulberry silkworms a) Eri silkworm, *Philosamia ricini* b) Tasar silkworm *Antheraea mylitta* c) Muga silkworm *A. assamensis*

The mulberry silkworm can be easily distinguished from non-mulberry silkworm by the following character.

Non-mulberry Silkworms

Cocoon

	Eri silkworm	Tasar silkworm	Muga silkworm
Host plants	Castor, Papaya Tapioca, Ailanthus	Terminalia, Dalbergia, Zizyphus ficur etc	Machilus, Bombycina zlitae polyantha, (Litsaea polya*nthus*)
Cocoon			
Colour	White or brick red	Brown	Lustrous golden yellow
Filament	Neither uniform or continuous	Uniform and continuous	Uniform and continuous
Silk Production	Moderate	High	Less

Important species exploited for tasar silk production

Tropical tasar	*Antheraea mylitta* = (A. paphia) (India)
Temperate tasar	*A. proylei* (India)
Chinese tasar	*A. pernyi* (China & Russia)
Japanese tasar	*A. yamamai* (Japan)

Some Non-commercial Sericigenous Insects

Moon moth	*Actias selene*
Cashew caterpillar	*Circula trifenestrata*

Commercial Silkworm

This is domesticated species and has been exploited for over 4000 years.

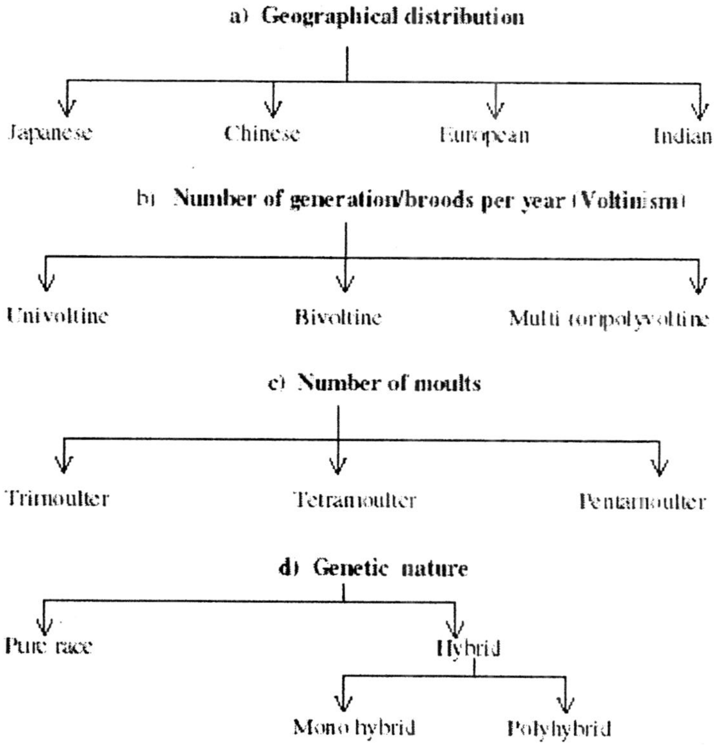

Classification of B.mori

All the different races of the present species have been derived from the wild ancestor, Mandarina silkworm, *B. mandarina*, (n = 27) which is a pest on mulberry in Himalayan region. The important commercially exploited South Indian races are:

Tamil Nadu White (TW)	—	Multivoltine, white cocoons with pointed ends
Pure Mysore (PM)	—	Multivoltine, one side of the cocoon broad and other pointed (egg shaped)

Hosa Mysore (HM)	—	Multivoltine capable of producing 30 35 kg/100 dfls (disease free layings)
Kalimpong A (KA), NB 7	—	Bivoltine race with oval white cocoons
NB_4D_2 & NB_{18}	—	Bivoltine race with peanut white cocoons
C. nichi	—	Bivoltine (or) multivoltine, cocoons groundnut shaped with a constriction in the middle (dumbbell shaped)
Nandi	—	A hybrid between KA x NB_4D_2. The constriction in the cocoon is very shallow
Chamundi	—	A hybrid between NB7 x NB18 capable of yielding 50 kg/100 dfls

The silk worm races are given below according to voltinism and genetic nature

Multivoltine	Bivoltine	Multi x Bivoltine cross breeds	Bivoltine hybrids
TW	KA	PM x NB4D2	KA x NB4D2
PM	NB4D2	TW x NB4D2	NB7 x NB18
HM	NB7	PM x NB18	NB18 x NB7
C. nichi	NB 18	RD1 x NB18	ND4D2 x KA
Nistari		BL23 x NB4D2	NB7 x NB4D2
Mysore 1		BL24 x NB4D2	CSR2 x CSR4
			CSR3 x CSR5
			CSR 18 x CSR 19

Morphology of *B. Mori*

Mulberry silkworm is a holometabolous insect and passes through four morphologically different stages in its life cycle viz., egg, larva, pupa and adult. Egg: Tiny, 1-1.3 mm length, 0.9 - 1.2 mm width ovoid, flat ellipsoid, oval and flat with micropyle at the anterior pole slightly off-centre. 2000 eggs weigh a gram. Races producing white cocoons lay pale yellow eggs while races producing yellow cocoons lay deep yellow eggs. In both the cases, the Japanese races lay slightly darker eggs than the Chinese races. In the case of hibernating eggs laid by bivoltine and univoltine races, the egg colour changes and becomes dark brown or purple with the deepening of the colour of the serosal pigments.

Larva

Newly hatched larva is black or dark brown in colour with large head, eyes, a pair of antennae and chewing type of mouth parts. Mouth has a median process, the spinneret through which the silk fluid from the silk gland is emitted forming the silk filament. Body is densely covered with bristles and looks like hairy caterpillar. Four pairs of tubercles viz., sub-dorsal, supra spiracular, infra spiracular and basal and each of which carried 3-6 setae. As the larva grows, it becomes smoother and lighter in colour due to the rapid stretching of the cuticular skin during the different instars of the larval stage

Thorax

It has three segments each bearing a pair of three jointed legs on the ventral side, legs are conical in shape with sharp claws at their ends to hold the leaf. On the dorsal side, middle segment possesses larval markings known as the eyespot. Abdomen: It is comprised of 11 body segments although only 9 can be distinguished as the last three are fused together to form the apparent ninth segment, the anal plate and the caudal plate. Crescent shaped marking (lunule) is present in 2nd abdominal segment.

Star like spot is present in 5th segment. Caudal horn is present on the dorsal side of eighth abdominal segment. Pair of flushy stumpy, unjointed muscular projection i.e., abdominal legs are present in abdominal segments 3rd, 4th, 5th & 6th. The abdominal segments carry the sexual marking which develop distinctly in the fourth and fifth instars in the eighth and ninth segments on the ventral side.

Female: sexual markings appear as a pair of milky white spots on each of the eighth and ninth segments. Pair of spots in eighth segment is known as ISHIWATA'S fore glands while in ninth segment is known as ISHIWATA'S HIND GLANDS.

Male: Small milky white body HEROLDS GLAND appears at the centre of the ventral side between the eighth and ninth segments.

Pupa

It is a non-motile and non feeding stage. The larval organs are destroyed and adult organs are differentiated during this stage.

Adult

The body is divided into three regions viz., head, thorax and abdomen. The male and female moths can be identified from the size of the abdomen, the size of antennae and the external genitalia.

Anatomy

The anatomy of the larva is different from that of the adult and the changes are brought about in the pupal stage.

Silk Glands

Silk glands are transformed labial glands, ectodermal in

origin, cylindrical and tubular with characteristically branched nuclei. These are situated in the ventro—lateral side of the mid intestine and the posterior ends are blind. Anterior ends of the paired ducts unite and open into the spinneret. Silk gland may be divided into 3 distinct regions viz., anterior, middle and posterior regions which have three layers. The layers are outer tunica propria (uniform thickness), middle glandular layer (varying thickness) and inner tunica intima (varying thickness).

Anterior region: It is a straight tube opening at the fore end into the duct and posteriorly into the middle region.

Middle region: It is the largest of the three regions and has 3 definite flexions. The middle region is again divided into three functionally different sections viz., anterior, middle and posterior sections. The beginning of the middle region is narrow but widens suddenly. The middle part is the widest and the end of the rear part is narrow.

Posterior region: It is crooked and curved between the dorso-visceral muscles and the tracheae.

A pair of glands known as FILLPPI'S or LYONNET'S gland is situated at the junction of the two anterior regions. These glands secrete a viscous fluid and their exact function is not clear. Fibroin and sericin are secreted in the fourth and fifth ages. Fibroin, the silk protein is secreted from the posterior region. The middle region acts as a reservoir for the maturation of fibroin and also secretes sericin I, II & III around the fibroin.

Sericin I: The innermost sericin is secreted from the posterior section of the middle region.

Sericin II: The middle layer sericin from the middle section of the middle region.

Sericin III: The outer most sericin from the anterior section of the middle region

The anterior region of the silk gland does not secrete any particular substance. It is simply a passage carries the silk substance from the reservoir (middle region) for freezing. The thread of the two sides is called brins and the sericin layer of two brins bind them together into single filament or bave.

Rearing

The most important principles to be remembered in silkworm rearing house are

- — Damp condition
- — Stagnation of air
- — Direct and strong drift of air
- — Exposure to bright sun light and radiation

Ensure

- — An equable temperature and humidity
- — Good ventilation

Rearing House

- — A separate house is ideal for rearing silkworm
- — The rearing house should have sufficient number of windows to permit cross ventilation.
- — Provision should be made to make it air tight for proper disinfection.
- — Draw the design of a model rearing house with a rearing capacity of 400 to 500 dfls at a time.

Brushing

Process of separating neonate larvae from their egg shell and transferring them to the rearing bed.

Grainage

An establishment where disease free and quality seeds (industrial seeds) are produced on scientific lines in large scale to be supplied to commercial rearers.

Preparation for Brushing

— Before commencement of each rearing, the rearing equipments and rearing houses must be thoroughly washed and disinfected with formalin.

— Two per cent formalin is sprayed on equipments, walls, roof and floor uniformly to destroy the disease causing organisms.

— The rooms should be kept closed for about 24 hours after disinfection.

— The doors and windows should be kept open at least for 24 hours before commencement of rearing to avoid traces of disinfectant.

— Different concentrations of formalin are used for applying to rearing room and appliances.

Purpose	% used
1. For routine disinfection	1 to 2
2. After pebrine infection	4 to 5

— The commercial formalin contains 35 to 40% formaldehyde with a specific gravity of 1.081 to 1.087 respectively.

— The required formalin concentration is prepared by diluting the commercial formalin by using the formula:

$$\text{Qty of formalin required} = \frac{\text{Required concentration} \times \text{Required qty of spray solution}}{\text{Available concentration of formalin}}$$

Normally 460 cc of formalin solution is required to spray an area of 10 m^2. Formalin has been reported to be carcinogenic. Hence, it is advisable to use bleaching powder 5%. Disinfection can be done three days before brushing. Use bleaching powder at 100 g/l around the rearing rooms daily @ 200 g/sq. m for disinfection and 60 g/l for washing the hands.

Appliances

Silkworms rearing appliances

— Rearing stand: Used for supporting the rearing trays. It is made up of locally available wood / bamboo. The standard sized stand has a height of 2.5 m, length of 1.5 m and a width of 0.65 m. This can accommodate 10-12 tiers of rearing trays.

— Rearing trays: A large number of silkworms are accommodated and fed in the rearing trays and these are placed one above the other in the rearing trays.

— Ant wells: These are placed below the legs of rearing stand and are filled with water to prevent ants from crawling up the stand and harming the silkworms.

— Paraffin paper: The craft paper coated with paraffin wax is used to prevent the evaporation of moisture and to maintain high humidity in the rearing bed of the young silk worm.

— Foam rubber strips: These are also used to maintain high humidity in the bed of young worms.

— Chop sticks: These are used like a forceps for picking worms, particularly the diseased worms away from the bed. These are two thin sticks of bamboo (20 cm), which taper to a point. At the thick end a thread connects the two sticks and looks like a forceps.

— Feather: Bird feathers (white ones) are used for brushing the newly hatched larvae from the egg card to the rearing trays.

Feeding appliances

— Leaf basket: Bamboo baskets of convenient size used for collecting and transporting the mulberry leaves from the field to rearing house. Wet gunny bags and polythene bags are also being used in place of bamboo baskets.

— Leaf chamber: It is used to store the harvested mulberry leaf. The sides and bottom of the chamber is made up of strips of wooden reepers. Wet gunny cloth is placed as a lining on all sides and top. Water is sprinkled periodically on the gunny cloth to keep the leaves fresh.

— Chopping board: Rectangular board made of soft wood used for cutting mulberry leaf to the desired size.

— Chopping knife: Sickle like knives with a broad and sharp blade are used for chopping mulberry leaves prior to feeding larvae. The knives should be of good quality and should not make any splits while cutting the leaves since which may increase rate of evaporation from the leaves.

— Mats: These are placed below the chopping board prior to chopping and used to collect cut leaves. Clean newspapers can also be used.

— Feeding stand: A folding stand on which the trays removed from the stand are placed, one at a time for feeding and bed cleaning.

Incubation of Eggs and Chawki Rearing

Transport of Eggs

— The disease free layings should be purchased from a recognised grainage.

— Purchase the eggs within four days after oviposition or at blue egg stage.

— While transporting the eggs from grainage to the rearing places, don't expose the eggs to sun light, wind, rain, etc.

— Transport the eggs in an egg carrying box/bag specifically designed for it.

— Transportation of the diapausing egg is recommended during the aestivation period.

— Eggs may be transported for short distances in the hibernation period.

— Transportation should be avoided in the preincubation period.

— During incubation, transportation should be avoided during torsion period (i.e.48-72 hours after laying).

— They may be transported during or after the maximum length period (i.e., on or after the fifth day of incubation).

Incubation and Hatching

— The egg sheets should be spread out as a single layer in a chawki tray.

— Temperature of 25°C and humidity of 80 per cent are maintained. For this, paraffin papers and wet foam pads may be used.

— When the eggs come to head pigmentation stage (about 48 hours before hatching), they should be kept in dark condition by wrapping them in a black paper or by keeping them in a box (black boxing). On the expected day of hatching, eggs are exposed to light, early in the morning to ensure uniform hatching. This facilities uniform development of embryo.

— Most of the eggs (90 to 95 per cent) will hatch in about 2 to 3 hours.

Preservation of Eggs

The eggs can be kept in an earthern incubation chamber.

Brushing

— The hatched larvae should not be starved and they must be brushed on a paraffin paper in a rearing tray or blue polythene film sheet (Rearing bed).

— This is done by sprinking chopped tender mulberry leaves of size 0.5 to 1 cm^2 over the hatched larvae. The larvae crawl on to the leaves.

— After 8 to 10 minutes, the egg sheet is inverted over rearing tray and gently tapped.

— Worms that are still attached to the egg sheets should be gently removed to the tray with a feather.

— A rearing bed is prepared and some more chopped leaves, if necessary, are sprinkled.

— To prevent drying of leaves and to maintain the required humidity in the rearing bed, wet foam pads and paraffin paper covering are provided.

Young Age Silkworm Rearing

In a tray of 120 x 90 x 105 cm size, 20 dfls are brushed

and reared upto second stage. Indicate the temperature and humidity requirement, bed spacing for every unit of 20 dfls for each day, number of feedings and quantity, cleaning, etc. for young age silkworm rearing.

Leaf Selection

- From brushing to the end of second age, the larvae are fed with tender leaves.
- The leaves are selected from the largest glossy leaf, 3rd or 4th from the top.
- The next 6 to 8 leaves are used to rear the young age worms upto II moult.
- The size of the chopped leaf is around 0.5 to 1.0 sq. cm. during 2nd age.
- Illustrate with the help of a figure, the selection of leaves from a fully grown branch.
- In the second age, two cleanings are given, one after resuming feeding and the other before second moult.
- A net with mesh size of 0.5 x 0.5 cm is spread over the rearing bed and feeding is given.
- The worm crawl through the net and come to fresh leaves.
- The net along with the worms and leaves are transferred to another tray.
- The left over leaves and litter are discarded.

Leaf Preservation

- Silkworm grow best when fed with succulent leaves which are rich in nutrients and moisture.
- The leaves, if not preserved properly, dry up and become unsuitable for feed.

— The harvested leaves must be preserved in fresh condition in a wet gunny cloth.

— If the climate is too hot and dry, the leaves are preserved in a leaf chamber which is lined with gunny cloth. The cloth is kept wet by spraying water at frequent intervals.

Cleaning

— It is the process of removing the silkworm excreta and left over leaves in the rearing bed.

— In the first age, one cleaning is given just a day before the worms settle for moulting.

— In the second age, two cleanings are, one after resuming feeding and the other before second moult.

— A net with mesh size of 0.5 x 0.5 cm is spread over the rearing bed and feeding is given.

— The worms crawl through the net and come to fresh leaves.

— The net alongwith the worms and leaves are transferred to another tray.

— The left over leaves and litter are discarded.

Moulting

— At the time of moulting, care should be taken not to disturb the worms.

— Correct detection of moult and stopping or resuming feeds are very important for uniform growth of silkworms.

— During moult, the rearing bed should be kept thin and dry and should have proper aeration.

Late Age Rearing

— The third, fourth and fifth instar larvae are considered as late age worms. They are reared in bamboo trays. Newspapers are spread over the trays to absorb excess moisture in leaves and faecal pellets.

— The temperature and humidity requirement gradually comes down as the stage advances.

— Leaves of medium maturity (16th leaf onwards) are fed in the third and fourth age and coarse leaves are fed in the fifth age.

— Over matured and yellow leaves should be rejected, since it may induce disease outbreak.

Rearing schedule of the late age worms (for 100 dfls)

Factors	III instar	IV instar	V instar
Temperature °C	26-27	25-26	23-25
Humidity (%)	75-80	70-75	65-70
Period (days)	3-4	4-5	6-7
Size of leaf (cm^2)	4-6	Entire leaf (or) chopped shoots	Entire leaf (or) chopped shoots
Total quantity of leaves (kg)	25-30	75-80	550-600
No. of feeds per day	4	4	4
No. of cleanings	3	Every day	Every day
Spacing (m^2)	4.18 – 8.36	8.36 – 16.72	16.72 – 33.44
No. of bamboo trays	5 – 10	10 – 25	25 – 50

Moulting

— Remove the paraffin papers

— Evenly spread the larvae in the rearing bed 6 - 8 h before settling for moult.

— Provide air circulation to avoid excess humidity inside the room.

— Provide charcoal stove / heaters to raise the room temperature during winter.

— Apply lime powder at 60 minutes before resumption of feeding daily during rainy / winter seasons to reduce the dampness in bamboo trays.

Mounting

— After attaining full growth in the final instar, the worms cease to feed and are ready to spin. Such worms are slightly translucent and raise their heads to find a place for spinning.

— These worms have to be picked up and transferred to a mountage for spinning cocoons.

— Mounting of worms should not be delayed as the ripened worms will waste silk.

— About 800-900 worms per m^2 are to be kept on a mountage. For 100 dfls, about 30 to 40 chandrikes are required.

— Mountages should be kept in shade in a well ventilated place during spinning.

Care During Spinning

— Quality of silk depends on the care taken at the time of spinning.

— Mature worms are sensitive to temperature, humidity, light, etc., at the time of spinning.

— The ripe worm require space equal in area to square of the length of its body for spinning.

— Proper spacing avoids wastage of silk for forming preliminary web and avoid double cocoons.

— To prevent staining of cocoons, keep mountage in an inclined position so that the urine may drop to the ground.

Humidity Maintenance

— Fluctuation of humidity causes abrupt thinning and thickening of silk filament.

— A relative humidity of 60 - 70% is ideal for spinning.

— Provide proper ventilation and straw mats below the mountage to absorb the liquid excreta.

— Provide even and moderate lighting. Improper lighting (bright light or dark shadow) causes crowding of larvae to shaded area leading to double cocoons.

— Remove dead worms and non-spinners on the 2nd day of spinning.

— To protect the silkworms from predatory ants, apply malathion 5% dust at the base of mountage stand.

Harvesting

— The silkworms complete spinning in 48 to 72 hours but the cocoons should not be harvested at this time as the worms inside are still in the prepupal stage.

— Harvesting should be done on the fifth day (7th day for bivoltine hybrids) when pupae are fully formed and hard.

— Do not harvest when the pupa is in amber colour.

— Dead and diseased worms on the mountages should be removed before harvest.

— Marketing of cocoons should be done on the sixth day (8th day for bivoltine hybrids).

Shoot Rearing

Silkworm larvae consume 80% of their food requirement during fifth instar. Fifty per cent of the labour input is utilised during the last seven days of rearing.

Rearing House

Provide separate rearing house for shoot rearing in shady areas. Separate room should be provided for young age worm rearing, leaf storing and hall for late age worm rearing.

Shoot Rearing Rack

— A rearing rack of 1.2 x 11 m size is sufficient to rear 50 dfls.

— Provide 15 cm border on all sides of the shelf to prevent the migration of the larvae.

— Fabricate the rack stand with wood, or steel and the rearing seat with wire mesh / bamboo mat.

Shoot Harvesting

— Cut the shoots, when they are at 1 m, shoots can be harvested 60-70 days after pruning, store the shoots vertically upwards in dark cooler room, provide thin layer of water (3 cm) in the storage room and place the cut ends in the water for moisture retention.

Feeding

— Provide a layer of newspaper in rearing shelf.

— Disinfect the bed, spread the shoot in perpendicular to width of the bed.

— Place top and bottom ends of the shoots alternatively to ensure equal mixing of different qualities of leaves.

— Transfer the fourth instar larvae after moulting.

— Watch for feeding rate from 4th day of fourth instar. If 90% of larvae have not settled for moulting, provide one or two extra feedings.

— Provide 3 feedings during rainy/winter months and 4 feedings during summer rearing.

Bed Cleaning

— Bed cleaning is done once during 2nd day of fifth instar following rope (or) net method.

— In rope method, spread 2 m length of rope (two numbers) at parallel row, leaving 0.5 m on other side.

— After 2 to 3 feedings, ends of the ropes are pulled to the centre to make it into a bundle.

— In net cleaning method, spread 1.5 cm^2 size net across the bed.

— After 2 or 3 feedings, the nets are lifted and the old bed is cleaned and disinfected.

— Transfer the net to newer shelf, spread the net over the shoots, larvae will migrate to lower layer.

Advantages

— Labour saving upto 70% when compared on hour to hour basis with leaf feeding method.

— Leaf saving upto 15-20%. Hence, leaf cocoon ratio is less by 2-3 kg and extra cocoon production.

— Better cocoon characters and effective rate of rearing (ERR).

— Better preservation of leaf quality both during storing and on the bed.
— More organic matter production (upto 18 onnes per ha per year).
— Better hygienic conditions can be maintained.
— Handling of silkworms minimised. Hence, contamination and spreading of disease reduced.
— Bed cleaning only once after IV moult.
— Worms and leaves are kept away from the litter. Hence, chances of secondary contamination are minimised.
— Labour dependent risk is reduced.

Disadvantages

— Required rearing room floor area is more (by 30%).
— Bed refusals will not be available as a cattle feed.
— Planting materials (cuttings) will not be available.

Pests of Silkworm

Uzi fly Exorista bombycis Ciedemann. Tachinidae; Diptera

Type of damage

Mature maggot causes reduction in yield of cocoons and cocoon quality. Causes death of silkworm larva.

Symptoms

— Presence of creamy white oval eggs on the skin of larvae in the initial stage.
— Presence of black scar on the larval skin.
— Silkworm larvae die before they reach the spinning stage (if they are attacked in the early stage).
— In later stage, pierced cocoon is noticed.

Period of occurrence

Throughout the year, severity is more in winter months.

Prevention and control

— Maintain sanitary and hygienic conditions in the rearing room.

— Provide physical barriers like wire mesh in the doors and windows of the rearing rooms.

— Dust laevigated china clay 3-4 gms per 100 spinning larvae and per sq. ft area of spinning tray.

— Spray 1 per cent benzoic acid over the larvae kill the eggs of uzi fly.

— Dissolve the uzicide tablets in the water (2 tablets/l) to attract the adults.

— Release the gregarious, ectopupal hyperparasitoid Nesolynx thymus Girault (Eulophidae: Hymenoptera) @ 1 lakh adults/100 dfls during night hours.

— Release the hyperparasitoid in three split doses @ 8000, 16,000 and 76,000/100 dfls during fourth and fifth instars and after cocoon harvest.

Dermestid Beetles

Dermestes ater, D. cadaverinus, D. valpinus, Anthrenus verbasi, A. pipinellae Dermestidae; Coleoptera

Type of damage

Grubs and adults reduce the quality of cocoons and damage the pupae kept for pairing in grainage.

Symptoms

Grubs and adults bore into the cocoons and eat the dried

pupae, attack pierced and melted cocoons stored within the grainage. Presence of small holes (Pierced cocoons) in the pupae, abdominal parts are damaged in the adult moths. Period of occurrence: Throughout the year in tropics.

Prevention and control

— Store the pierced cocoons in a separate room.

— Avoid long storage of pierced cocoons.

— Maintain hygienic conditions in the rearing house and cocoon storage rooms.

— Sun dry the pierced cocoons once in a week. Dip the wooden utensils in 0.2 per cent malathion for 2-3 minutes. After 10 days, wash the trays in water and sun dry and reuse them.

Straw Itch Mite

Pediculoides ventricosus (New port), Pediculoididae

Type of damage

Nymphs and adult mites cause reduction in cocoon yield.

Symptoms

Larvae, pupae and adult silkworms are damaged. Larvae stop feeding, become sluggish, body turn to purple brown, vomiting with yellowish brown fluid, beaded faeces attached to the anus. Presence of uneven black spots on larval skin. Pupae turn to black and develop lesions.

Period of occurrence

May to middle of September.

Diseases of mulberry silkworm and their management

Virus Diseases

S. No.	Description	Nuclearpolyhedrosis/ grasserie	Cytoplasmic polyhedrosis virus (CPV)	Infections Flacherie virus IFV	Densonucleosis virus
1.	Causal agent	Borrelina bombycis virus, Baculoviridae	Smithia virus, Reoviridae	Morator virus Picornaviridae	Denso virus, Parvoviridae
2.	Site of infection	Nuclei of tracheae, fat bodies, epidermis, haemolymph	Cytoplasm of cylindrical cells of mid gut	Cytoplasm of globet cells of midgut	Nuclei of cylind-rical cells of midgut
3.	Shape of polyhedra	Hexahedron DNA	Irregular hexahe-dron tetragon	No polyhedra formed	No polyhedra formed
4.	Symptoms	Body colour-milky white, swollen segments(bamboo cane like), hang up side down, body wall burst on slight	Milky white tinge in advanced stage, diarrhoea, motio-, nless irregular milk white in posterior part	Diarrohaea, motionless, irregular shaped faeces, brownish gut, body transparent	Diarrohaea, motionless, irregular shaped faeces, brownish gut,does not burst

	disturbance, haemolymph-milky white with numerous polyhedral bodies.	of midgut, annular wrinkles	burst uneasily	transparent
5.Source of infection	Contaminated ,leaves and environment peros or wound infection	Contaminated leaves and environment perosor wound infection Faeces Peros	Faeces Peros	Faeces Peros
6. Diagnosis	Polyhedra in haemolymph	Polyhedra in the posterior part of midgut	Biological and serological test	Biological and serological test

S No.	Description	Septicaemia	Sotto
1.	Causal agent	Black thorax septicaemia Bacillus sp. Bacillaceae Red septicaemia by Seratia marcescens	Strains of Bacillus thuringiensis, Bacillaceae
2.	Site of infection	Haemolymph	Gut and nervous system
3.	Characters of spores	Spores subterminal, gram positive in Bacillus sp.S.marcescens is non-sporulating and gram negative	The spore produce delta endotoxin
4.	Symptoms	Common disease symptoms are sluggish movement, decreased appetite, straightened body, swollen thorax, shrinkage of abdominal segments.Vomitting and bead loss like faeces, of clasping power of prolegs. Body becomes soft and discoloured. Body wall easily ruptures and emits foul smelling fluid	Loss of appetite, convulsions, lifting of head,spasm, tremors, paralysis, distress, sudden collapse and death. Shortly after death, the crop is outstretched, hard to touch and the head appears hook shaped. Body turns to brown,blackish brown and black and rot exuding foul smelling dark brown fluid.

5. Pathological changes	Body wall ruptures easily emitting foul smelling fluid	Body is outstretched, hard to touch and headappears hook shaped.Exude foul smelling dark brown fluid.
6. Source of infection	Injury or wounds and rarely orally	Usually peroral and wounds / injury.
7. Diagnosis	By differential staining technique	By differential straining technique
8. Course of the disease	Acute	Acute

Fungal Diseases

S.. No	Description	White muscardine	Green muscardine	Aspergillosis
1.	Causal agent	Beauveria bassiana (Balsamo), Moniliaceae	Nomuraea rileyi Farloue, Moniliaceae	Aspergillus flavus Link, A.oryzaeehmer,Moniliaceae
2.	Site of infection	Body wall, spiracles	Body wall and haemolymph	Body wall
3.	Characters of conidium	Colourless, globular or oval shape and porcelain white	Conidium oval shaped and slightly pointed at one end Light green and single celled	Conidia spherical
4.	Symptoms	Appearance of moist specks on the skin, loss of appetite and inactive. Body becomes limp, of skin elasticity, stops movement loss diarrhoea and vomitting and finally dies	Appearance of dark brown irregular lesions on the ventral, dorsal and lateral and sides of diseased larvae, diarrhoea, vomitting and death, whole body is covered with mycelia with fresh green conidia.	Infects mainly first and second instar. Larvae cease eating, become lethargic, show body tension, lustrousness and then die. Just before death, the head and thorax extended outwards

			and vomiting occurs. Body is covered with conidia
5. Pathological changes	Body becomes stiff and hard, body colour changes to ened pink colour, wolly aerial hyphae grow out between intersegmental membrane. Body is covered with white powdery conidia, cocoon sound like dried cocoons when shaken. Body of moth is hardened and the wings fall off easily.	Body wall becomes hard- and covered with mycelial mat with green conidia	Hardening of corpse in dead larvae at the site of fungus penetration and other parts become black and rotten
6.Source of infection	Mummified larvae, infected seat paper, tray and dead wild lepidopterous larvae from the mulberry field.	Dead mummified larvae and infected wild lepidopterous insects from field	Mouldy objects and high humidity
7.Diagnosis	Conidial examination under microscope.	Conidial examination under microscope	Conidial examination under microscope

Management

— Disinfection of the rearing rooms and trays.

— Avoid keeping cotton, wheat straw or rice straw near the rearing rooms/appliances.

— Pre-rearing disinfection with sulphur @ 30 gm/ Cuft. Besides the above pests, ants, rat, earwig and nematode attack the silkworm and stored cocoons also.

Prevention and control for viral diseases

— Sun drying of rearing appliances for one/two days.

— Disinfection of rearing room and appliances with 3% bleaching powder.

— Disinfection of worms, trays and discarding of diseased worms.

— Ensure proper ventilation and air circulation.

— Early diagnosis and rejection of infected lots.

— Provide proper bed spacing.

— Collect and burn infected larvae, faecal matter and refuges.

— Feed the larvae with nutritious mulberry leaves.

— Dust the bed disinfectant, Vijetha (or) Reshamkeet Oushadh on the larvae, ½ hr after each moult before resumption of feeding (3 kg/100 dfl).

— Spray 1% extract of Psoralea coryleifolia on mulberry leaves, shade dry and fed the worms once during third instar.

Prevention and control for bacterial diseases

— Disinfection of rearing room and appliances

- Disinfection of worms, trays and discarding of sick worms
- Maintenance of hygienic condition
- Feeding the larvae with nutritious leaves
- Early diagnosis and rejection of infected lots
  - Avoid injury to the worms, overcrowding of trays and accumulation of faeces in the rearing bed
  - Sound management, improving the rearing environment and feed stuff
  - Apply antibiotics like Streptomycin/Tetracyclin / Ampicillin on leaves.

Prevention and control for fungal diseases

- Sundry the rearing appliances
- Disinfect the rearing room and utensils with 2 per cent formalin or 5 per cent bleaching powder
- Avoid low temperature and high humidity in the rearing room
- Keep the rearing bed thin and dry
- Early diagnosis and rejection of infected lots
- Apply Dithane M45 on the larvae (3 kg/100 dfls)
- Disinfection of rearing rooms and trays with 4 per cent pentachlorophenol (for Aspergillosis)

Protozoan Disease

Pebrine

Causal Agent	Nosema bombycis, Nosematidae
Causal Organism	Protozoan
Site of infection	Gut and cytoplasmic diseases

Characters of spores	Mature spore is oval with three layered membrane and appears light blue under microscope.
Symptoms	Observed in eggs, larvae, pupae and adults
Egg Stage	Poor egg number Lack of adequate adherence to the substratum Lack of egg uniformity, more of unfertilised and dead eggs Poor and irregular hatching
Larval Stage	Poor appetite, retarded growth, irregular size, irregular moulting and sluggishness. Transovarially infected larvae die before third moult. Body wrinkled with rustic brown colour. Irregular pepper like spots on the larval skin.
Pupal Stage	Pupae are flabby and swollen with lustless and softened abdomen. Pupae fail to metamorphose into adults
Adult stage	Moth emergence is delayed, clubbed wings, distorted antennae, mating improper; scales easily come off. Moth lays eggs with less gluey substance.
Course of the disease	Chronic
Pathological Changes irregular	Larval body shows wrinkled skin with rustic brown colour, pepper like spots on larval skin, Pupae are flabby and swollen. In

	adults, presence of clubbed wings and distorted antennae.
Source of infection	Rearing of transovarially and surface contaminated layings, diseased and dead larvae, faeces of larvae, moths, diseased egg shells, larval and pupal exuviae and contaminated mulberry leaf
Diagnosis	Examination of spores under microscope

Prevention and Control

— Produce healthy eggs

— Disinfection of rearing room and utensils

— Maintain strict hygienic conditions during rearing

— Surface disinfect the layings in 2 per cent formalin for 10 minutes before incubation

— Collect and burn the diseased eggs, larvae, pupae and adult moth, bed refuges, faecal pellets, etc.

2

Silkworm Genomics

During the past decade much effort has been made in the areas of marker development, and molecular maps have been constructed in standard strains with the use of RFLPs, RAPDs, ISSRs, STSs, and microsatellites. The potential applications of molecular markers and linkage maps include stock identification, Marker Assisted Selection (MAS), identification of Quantitative Trait Loci (QTL), and, ultimately, positional cloning of visible mutations and QTL. To these ends, BAC libraries have been constructed and are being used to make large-scale physical maps, with markers based on ESTs as framework anchors. Altogether this work provides a foundation for identification of gene function, gene and chromosome evolution, and comparative genomics.

The silkworm, *Bombyx mori,* domesticateda for silk production for about 5000 years is the most well-studied lepidopteran model system because of its rich repertoire of well characterised mutations affecting virtually every aspect of the organism's morphology, development, and behaviour and its considerable economic importance. By virtue of a long history of silkworm rearing for commercial purpose, silkworm genetics has been the subject of considerable research interest resulting in careful collection, cataloguing and maintenance of various silkworm genetic stocks of considerable scientific and economic interest.

Today, the opportunities for genetic manipulation and study of the silkworm, *B. mori* include more than 400 visible mutations out of which 200 have been assigned to conventional linkage groups covering 900.2 cM. These mutations affect many fundamental aspects of the insect's life cycle, including egg and egg shell formation, early embryonic development and pattern formation, larval feeding behaviour, molting, embryonic diapause, etc.

In addition, a vast array of distinct geographical races and inbred lines that represent variation for a number of qualitative and quantitative traits of basic biological and economic interest including body size, silk quality, fecundity, pathogen resistance, and heat tolerance are available. In fact, the silkworm *B. mori* is genetically the best known insect next only to the fruitfly, *Drosophila melanogaster*. The haploid genome size of *B. mori* is estimated to be 530 Mb, 3.8 times that of *D. melanogaster* and one-sixth the size of the mammalian genome.

Supported by the infrastructure of data and genetic resources available for research, the silkworm is a key model organism in Lepidoptera which includes more than 160,000 species, of which Bombycoid moths include silkmoths of economic importance and Noctuid moths, the largest group in the order and includes some of the most devastating pests of agriculture, particularly *Heliothis* (*H. virescens, H. armigera, H. zea*). The genomic information of the model species *B. mori* should be applicable to the most important species in Lepidoptera. Comparative studies on genome sequences between Lepidoptera and other published genome sequences could uncover lepidoptera-specific genes. Products of such novel genes could serve as targets for lepidopteranspecific insecticides.

Bombycoid moths secrete diverse varieties of silk fibres. These species include *B. mori* of family Bombycidae and wild silkmoths that belong to Saturniidae, *Antheraea mylitta* (Indian tropical tasar silkmoth), *A. pernyi* (Chinese oak tasar silkmoth), *A. assama* (Indian golden silkmoth), *A. yamamai* (Japanese oak silkmoth) and *Philosamia cynthia ricini* (Indian castor silkmoth). Silk production based on these moths, specially *B. mori, A. pernyi, A. mylitta* and *A. assama* plays important role in rural economies of many populous developing nations.

Six million people in India alone are involved in sericulture, which involves intensive labour and provides the key to improving local quality of life. In order to make sericulture economically viable, genes affecting growth rate, yield, fibre quality, virus resistance can be tagged with molecular markers for rapid construction of genetically improved strains. Considering the unique experimental advantages of this organism and its economic importance, an International Consortium on Lepidopteran Genomics was recently formed to promote international cooperation to sequence the genome of *B. mori* and to undertake comparative genomics of other economically important Lepidoptera.

Such international cooperation is expected to foster knowledge both in the basic and applied aspects of insect science. The current efforts of several laboratories contributing to the silkworm genome project have, as their immediate objective, the generation of high density molecular linkage map, development of tools and techniques to 'fingerprint' silkworm varieties, identification of lepidoptera-specific genes, and generation of a clone-based physical map that will allow virtually immediate access to any defined region of the euchromatin. The long-term objective is to identify the position, DNA sequence and function of all the genes in the organism. These are achievable goals considering the revolutionary developments in analytical tools and technical feasibility of large-scale DNA sequencing as demonstrated recently in complex genomes such as *Drosophila* and human.

Genetic Markers

In the past decade, several key advances in molecular biology have greatly advanced the impact of molecular genetics in biology. Most important have been: (i) development of the polymerase chain reaction (PCR), which amplifies specific stretches of DNA to usable concentrations, (ii) the application of evolutionarily conserved sequences as PCR primers, (iii) the advent of hypervariable microsatellite loci as genetic markers and (iv) the advent of routine DNA sequencing in biology laboratories. These innovations, coupled with the recent explosion of powerful analysis and relatively user-friendly computer programs, have accelerated the pace at which molecular genetic data can be used for various programmes.

These developments led to the detection, analysis and exploitation of naturally occurring DNA sequence polymorphisms in eukaryotes and laid a foundation for contemporary genomics of several organisms. Polymorphic genetic markers have found potential applications in animal and plant improvement programmes as a means for varietal and parentage identification, construction of molecular linkage maps and evaluation of polymorphic genetic loci affecting quantitative economic traits.

A wide array of DNA marker techniques is available for genetic studies. All DNA markers reflect differences in DNA sequences. Typically, in a diploid organism, each individual can have one or two different states (alleles) per character (locus). The choice of a particular genetic marker often depends upon the purpose of the study and is usually a trade-off between practicality and precision of genetic markers. One manifestation of this is the dichotomy between multilocus DNA techniques which include Random Amplified Polymorphic DNAs (RAPDs), Amplified Fragment Length Polymorphisms (AFLPs) and Inter-Simple Sequence Repeat PCR markers (ISSR-PCR) and single locus techniques which include single copy nuclear DNA regions and microsatellites.

Multilocus methods result in simultaneous visualisation of many dominant markers and are technically convenient, but have some marked weaknesses and limitations, including dominant inheritance – the DNA fragment can be scored only as present or absent, and in many cases substantial proportion of variation they detect can be non-heritable or not even derived from the target organism. By contrast, single locus markers are quite informative because they provide alleles whose zygosity status can be assayed easily.

Development of molecular markers is important in the silkworm for construction of linkage map, fingerprinting of strains for breeding, and marker-assisted selection. In the earlier studies, fingerprinting analyses of silkworm strains have been carried out using RAPDs, heterologous minisatellite probes, Bkm(2), M13, and ISSR-PCR. Genetic linkage maps of silkworm have been constructed using RAPDs, RFLPs, and AFLPs. Most of these markers, except RFLPs, are dominant and strain-specific.

Considering the genetic informativeness offered by microsatellite loci, which are co-dominant, a microsatellite discovery programme in the silkworm genome was initiated. Microsatellites, also known as simple sequence repeats (SSRs), are short stretches of nucleotides in which a motif of one to six bases is tandemly repeated, have emerged as ideal markers in eukaryotic genetics. They are ubiquitous in prokaryotic and eukaryotic genomes and are randomly distributed, both in protein coding and non-coding regions. The sequences in microsatellite motifs can differ in repeat number among individuals, providing a ready source of polymorphism.

With the advent of the PCR, this property of microsatellite DNA was converted into highly versatile genetic markers. PCR products of different lengths can be amplified using primers flanking the variable microsatellite region (Figure 1). Due to the advantages offered by microsatellites they have been used to construct linkage maps. In addition to their clear utility for practical applications, information about the distribution and variability of microsatellite sequences in the genome of a given species can help elucidate its genetic history from the standpoint of evolution and artificial selection.

Microsatellite Motifs

Analysis of microsatellite loci cloned and characterised from a *B. mori* sub-genomic library was reported. The total fraction of the genome analysed (2810 kb), which accounts for 0.53% of the *Bombyx* genome (530 Mb)

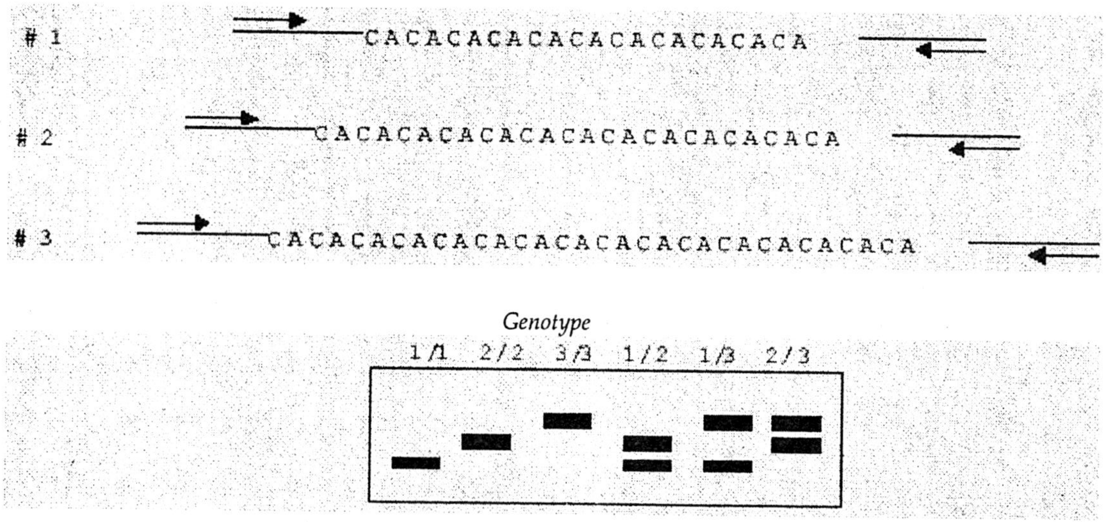

Figure 1. Microsatellite alleles and their detection. Three alleles of different sizes at a microsatellite locus composed of a (CA) repeat are shown. The arrows indicate the locus-specific primers designed for the region flanking the (CA) repeat, used for PCR amplification.

yielded 57 $(CA)_n/(GT)_n$ and 27 $(GA)_n/(CT)_n$ harbouring clones. The frequency of occurrence of dinucleotide repeats, was estimated to be, on average, one $(GT)_n$ every 49 kb, and one $(CT)_n$ microsatellite every 104 kb. The frequencies of occurrence of these two types of microsatellite motifs compare well with those reported in honey bee, pig, rat and human.

The perfect (no interruption in the run of repeats), imperfect (interruption in the run of repeats) and compound (a mixture of repeats of two or more motifs) repeat microsatellites in the silkworm genome were on the order of 62%, 29% and 7% respectively (Table 1). This is in close agreement with 72% perfect, 28% imperfect, and 11% compound repeats in human (Table 1).

Table 1. Percentage of different categories of (GT)n and (CT)n microsatellites in silkworm, honey bee and human. Number in parenthesis is the number of $(GT)_n$ or $(CT)_n$ microsatellites analysed

	Silkworm		*Honey bee*	*Human*	
Type of motif	*GT (21)*	*CT (7)*	*GT (23)*	*CT (52)*	*GT (57)*
Perfect	62	71	48	46	61
Imperfect	29	29	22	31	27
Compound	9	0	30	23	2

The primer pairs developed for the flanking sequences of the microsatellite loci isolated from the silkworm genome successfully amplified the repeats producing a variety variety of co-dominant markers. This indicated that the microsatellite loci were conserved and not generally derived from transposable elements, which are expected to be mobile, giving rise to null alleles. A summary of the microsatellite loci and PCR amplification conditions for sets of representative microsatellite loci are given in Table

Table 2. Microsatellite loci, repeat pattern, forward and reverse flanking primer sequences, number of alleles, allelic range, heterozygosity values and annealing temperature (T) in Bombyx mori

Locus symbol	Repeat motif length	Primer sequence: Forward	Primer sequence: Reverse	No. alleles	Allele size range (bp)	Hetero-zygo-sity	T (°C)	MgCl2 (mM)
Sat158	$(CA)_9$ $(CA)_4$	5′-cttccagacaaccattagat-3′	5′-cagcatccatc-cttatttat-3′	11	190-246	0.76	49	1
Sat211	$(GT)_{10}$	5′-gatcgaaactacgcattacg-3′	5′-cagcatccatc-cttatttat-3′	6	106-146	0.89	49	2
Sat256	$(CA)_5(TA)_7$	5′-attgcctcggacaggagtgt-3′	5′-tgtgaggaca-ggccgtta-3′	7	111-194	0.82	53	3
Sat346	$(GA)_9$	5′-gaagacagagcgaagtgga-3′	5′-atggattcctgc-tggtagat-3′	6	139-189	0.83	53	3
Sat892	$(GT)_{10}$	5′-caataaatgcttacgagtttaa-3′	5′-tatcggtagttc-cttgactt-3′	3	175-187	0.66	47	3
Sat951	$(GA)_{23}$	5′-attgtaaccgatttgagaga-3′	5′-attcgcacaataag-ttcact-3′	5	107-122	0.68	48	1
Sat962	$(GT)_{10}$	5′-cataataaataaataataat-3′	5′-tgtagtgggatag-tcagtat-3′	6	108-114	0.78	42	3
Sat1013	$(GT)_9$	5′-aacagatgctgcggactggt-3′	5′-tgccattcaca-atacaacat-3′	5	135-162	0.80	50	1

Contd....

Contd....

Sat1411	$(GT)_8(GT)_5$	5′-gaatgttctgctggtgg-3′5′-taatgttttatacttta-ttatatg-3′	8	109-162	0.68	45	3
Sat1423	$(CA)_{11}$	5′-ctttcgatcaccgcgttctc-3′5′-cgctacgaaatacca-ttatctgaca-3′	9	130-176	0.82	55	2
Sat1893	$(CA)_{10}$	5′-aatgcagaatcgtaatttt-3′5′-tttgaccacagaca-ataagc-3′	7	98-158	0.85	45	2
Sat2550	$(GA)_{12}$	5′-ggtcccttgaaactgcgat-3′5′-cagagacctgccgg-ttgtcgttc-3′	9	121-168	0.87	53	2
Sat2604	$(CT)_9$	5′-gctcgccatatgcaatcctc-3′5′-cgtcattgccttcattt-cagttc-3′	8	143-186	0.86	53	2
Sat2763	$(GT)_{22}(GT)_7$	5′-acgcgctctacaaaataccatta-3′5′-gatcaccggtt-ctgctctcg-3′	17	105-179	0.90	53	2
Sat3513	$(GT)_5$ (TGCG)$_6$(TA)$_2$	5′-cgcaattctgtattagataa-3′5′-taaaggtattattc-ttattcg-3′	6	134-223	0.84	46	3

2. The polymorphisms, which manifested as size differences in the PCR products, could be easily scored across silkworm strains.

Table 3. A summary of each type of marker assay performed on 13 silkworm strains

	Total number of assays	*Total no. products*	*Total poly-morphic products*	*Mean no. products per assay*	*% of polymorphic products*
SSR	15 (primer pairs)	122	105	8.10	86.00
Inter-SSR	6 (primers)	239	184	39.83	76.98
RAPD	40 (primers)	216	204	5.40	94.44
RFLP	6 (probes)	180	176	30.00	97.77

For example, the 15 microsatellite loci, 11 $(GT)_n$ and 4 $(CA)_n$, revealed a total of 113 alleles in 13 silkworm strains. The number of alleles scored at each locus in these strains varied from as few as 3 to as many as 17. The $(CA)_n$ motifs showed a tendency to reveal a greater number of alleles compared to $(GA)_n$ motifs. For example, 11 $(GT)_n$ loci revealed a total of 85 alleles, ranging from 3 to 17, with an average of 7.7 alleles per locus whereas the four $(CT)_n$ loci revealed a total of 28 alleles ranging from 5 to 9 with an average of 7 alleles per locus.

Heterozygosity Values

All 15 microsatellite loci revealed allelic length polymorphism in the 13 silkworm strains analysed. The average heterozygosity value was 0.79, ranging from 0.66 to 0.90. The most polymorphic locus, *sat2763*, which revealed 17 alleles, also showed the highest heterozygosity of 0.90. This locus harboured two stretches of (GT) motifs, one with 22 repeats and another with 7 repeats. The locus *sat892*, with a $(GT)_{10}$ motif that revealed only 3 alleles in the 13 silkworm populations, showed the lowest heterozygosity values. There was no clear evidence with regard to the relationship between the repeat length and degree of polymorphisms. For example, locus *sat951* (GA), revealed 5 alleles with a heterozygosity value of 0.68, and the locus *sat2604* with $(GA)_9$ showed 8 alleles with a heterozygosity value of 0.86. The loci that were similar in length also differed in their number of alleles and heterozygosity. The loci *sat211*, *sat892*, *sat962* and *sat1893*, which carry repeat motifs of $(GT)_{10}$, showed 6, 3, 6, and 7 alleles respectively. These results are in contrast to the reports on human and other species where longer repeats have been reported to generate a higher degree of polymorphism.

Genetic Diversity Studies

A variety of molecular techniques have been employed successfully in silkworm to detect genome-wide polymorphisms. These methods include Restriction Fragment Length Polymorphisms and a variety of PCR-based techniques. In a study reported by Nagaraju *et al.* and Sharma *et al.*, a banded krait minisatellite, Bkm(2)8 containing a 545 bp DNA sequence consisting of 66 copies of GATA repeats interspersed with a variable number of dinucleotide (TA) repeats in several locations, was used as a probe. The probe revealed intra- and inter-population genetic diversity when tested on 13 divergent silkworm genotypes.

The phenetic analysis of RFLP data separated the 13 strains into three groups: one consisted of all diapausing strains (except HU_{204}); the second consisted of all non-diapausing strains (except Pure Mysore and Nistari); and the third consisted exclusively of Pure Mysore and Nistari (Figure 2). In another study, six anonymous multilocus probes obtained from a *Pst*I subgenomic silkworm library, were used on the same set of 13 silkworm strains.

The genetic informativeness expressed in terms of number of loci revealed (effective multiplex ratio, EMR) and amount of polymorphism detected (diversity index, DI) was compared with the three PCR-based techniques, Random Amplified Polymorphic DNA (RAPD), Inter-Simple Sequence Repeat PCR (ISSRPCR) and simple sequence repeats (SSRs). The six multilocus RFLP probes produced 180 products of which 97% were polymorphic; the 15 SSR loci gave rise to an average of 8 alleles each, of which 86% were polymorphic; and ISSR-PCR produced 39 fragments of which 77% were polymorphic (Table 3). The highest diversity index was observed for ISSR-PCR (0.957) and the lowest (0.744) for RAPDs (Table 4). The RAPDs, ISSR-PCR and multilocus RFLPs clearly separated the diapausing and non-diapausing silkworm varieties, as did the Bkm(2)8 probe (Figure 2).

Table 4. A summary of mean DI, EMR and MI for different marker assays of silkworms

Marker assay	*Mean diversity index (DI)*	*Mean effective multiplex ratio (EMR)*	*Marker index (MI)*
SSR	0.795	1.0	0.796
Inter-SSR	0.957	23.6	22.570
RAPD	0.744	4.8	3.580
RFLP	0.930	28.7	26.670

The analysis of the same set of 13 silkworm strains using 5 different marker systems highlighted the utility of each of the marker systems in genetic analysis of silkworm. These results show that ISSR-PCR and multilocus RFLP markers are powerful techniques for fingerprinting silkworm varieties. Although most ISSR loci are dominant rather than codominant, ISSR-PCR markers offer several

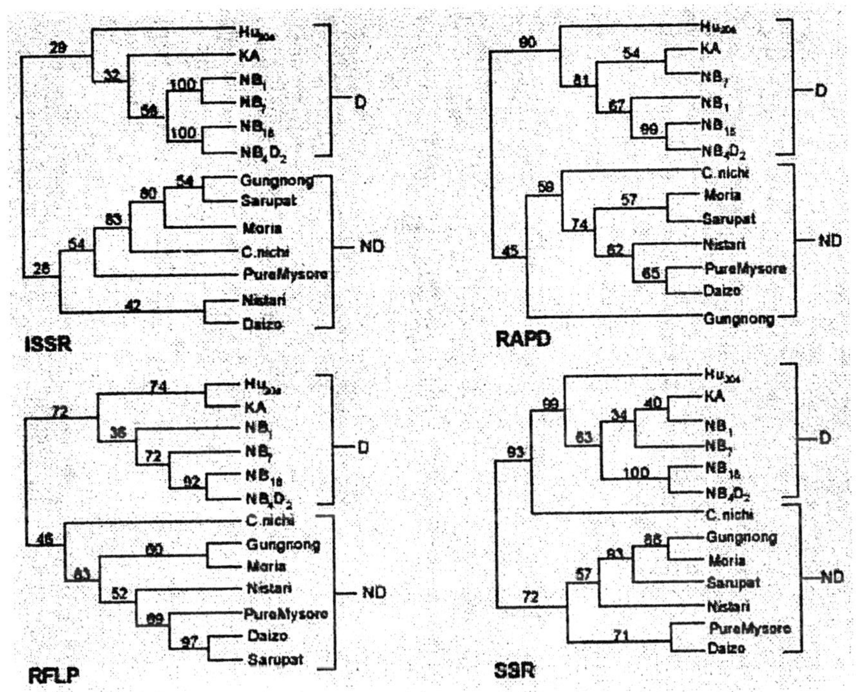

Figure 3. Dendrograms generated from genetic similarity matrices calculated using Nei and Li coefficients for ISSR-PCR, RFLP, RAPD and SSR marker assays. D, diapausing silkworm strains; ND, nondiapausing silkworm strains. Numbers on the nodes indicate the number of times a particular branch was recorded per 100 bootstrap replications following 1000 replications

advantages over RFLPs for genotyping, the major one being the rapid production of a large number of markers in a cost-effective manner. Because of these advantages, the ISSR-PCR technique has potential use in silkworm breeding, germplasm evaluation, and genetic mapping.

The recently automated ISSR-PCR technique by including a fluorescent dye in the PCR reaction followed by analysis on an ABI automated sequencer has been shown to be ideal for high-resolution mapping experiments and DNA profiling. Although multilocus RFLP probes offer features such as high reproducibility and easy scoring, the requirement of a large quantity of template DNA, and Southern hybridisation render them less suitable for large-scale studies. In the RAPD technique, though widely used, many factors such as concentration of Mg^{2+}, quality of the template, thermal cycler and the source of polymerase have been reported to affect the repeatability of the results between different laboratories.

In this aspect, RAPD analysis may not be a method of choice for large-scale genotyping experiments. Considering the reproducibility, genetic informativeness and speed, SSRs are the most useful markers among all the marker systems. Since SSRs are co-dominantly inherited, multiallelic in nature, and can be multiplexed and automated, they are useful for genetic diversity, pedigree evaluation and genetic mapping studies. Although the initial cost to develop locus-specific primers is high, the advantages that they are robust and are amenable to automation ultimately increase the cost benefit ratio, particularly considering the value of being able to compare data sets directly.

Cross-amplification of Bombyx *microsatellites*

Advances in lepidopteran genomics have been limited by

general lack of genetic information for individual species because of lack of genetic tools with broad applicability across species. An accessible approach to such a problem is to look for chromosome conservation across species by using comparative genome mapping. Recent results show that many *Bombyx* microsatellite loci are conserved across Saturniid silkmoths which include most of the commercially important wild silkmoths such as *Antheraea assama* (Muga silkworm), *A. mylitta* (Tropical tasar silkworm), and *A. pernyi* (Temperate tasar silkworm), as well as a Noctuid moth, *Helicoverpa armigera,* a serious agricultural pest. Considering that the genetics of these moths has hardly been studied, the microsatellite development programme in *B. mori* makes it possible to carry out molecular genetic analyses of these moths without investing in a microsatellite development programme for each species. The use of heterologous primers will significantly reduce the cost of developing markers for the wild silkmoths and agricultural pest, *Heliothis* spp. Besides, as mapping is already underway in *Heliothis,* a comparative mapping approach will enable the prediction of the presence of microsatellites associated with genes of interest. Such conserved loci could be potentially used for assessing the evolutionary relationships within and across species with a reasonable degree of accuracy.

Linkage Maps

The first attempt to construct molecular linkage maps for the silkworm utilised RFLPs generated by 15 characterised single copy sequences, 36 anonymous sequences derived from a follicular cDNA library, and 10 loci corresponding to a low copy number retrotransposon, *mag*. Recently, this work was extended using as probes partially sequenced cDNAs (ESTs) derived from an embryonic and a maternal mRNA library. By using the classic backcross strategy to detect linkage, a small number of segregants (e.g., 15 animals) from backcrosses of single heterozygous females derived from strains p50 and J02 to inbred J02-derived males have been used to assign more than 200 codominant markers to 27 linkage groups and one independent marker. Integration of these linkage groups with the conventional genetic maps of morphological and biochemical markers is under-way, together with the reciprocal backcrosses designed to assign recombination distances within each linkage group.

An RAPD linkage map of silkworm has been constructed using 320 random primers, which produced 243 polymorphic products between two parental silkworm strains, C108 and P50. In the F_2 population of 101 individuals used for these maps, segregation ratios of 168 bands were nearly 3 : 1 in chi square tests, showing Mendelian inheritance. The MAPMAKER program sorted 168 bands into 29 linkage groups. The sum of map distances is approximately 900 cM. Recently, Nagarajahas constructed a Z-chromosome RAPD linkage map using a backcross population (BC) derived from crossing an F_1 (male) heterozygous for a recessive sex-linked marker, transluscent larval skin (*od*) × a homozygous translucent (*od*) (female).

In silkworm, both F_2 (Figure 3) and backcross populations (Figure 4) have been used by different groups. Four groups have used an F_2 population derived from a cross of the same two reference parental strains, C108 and P50. Although both strains are of Chinese origin, they exhibit high phenotypic diversity for such complex characters such as size, growth rate, diapause, nutritional requirements, general vigour and cocoon properties such as cocoon weight, cocoon shell weight, and silk fibre length, suggesting that considerable polymorphism exists at DNA level.

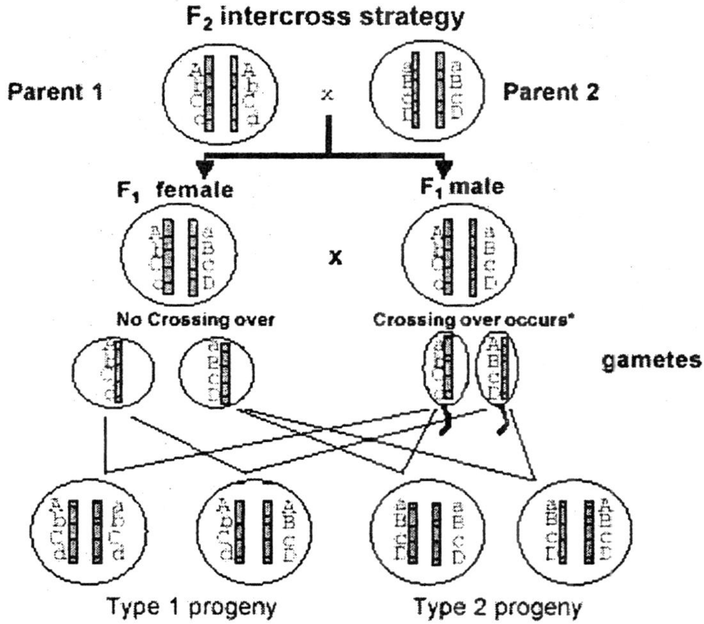

*Non-recombinant gametes not shown

Figure 3. Schematic representation of the behaviour of markers in an F_2 intercross

Different groups have used a different number of F_2 offspring for mapping, affecting the ability to detect linkage between distal markers. Shi *et al.* have used 52 F_2 progeny for RFLP maps, whereas Promboon *et al.*and Yasukochihave used 101 and 166 F_2 progeny, respectively, for maps based on RAPDs.

The SSR and Fluorescent ISSR (FISSR) markers have been mapped recently using the same 101 F_2 offspring

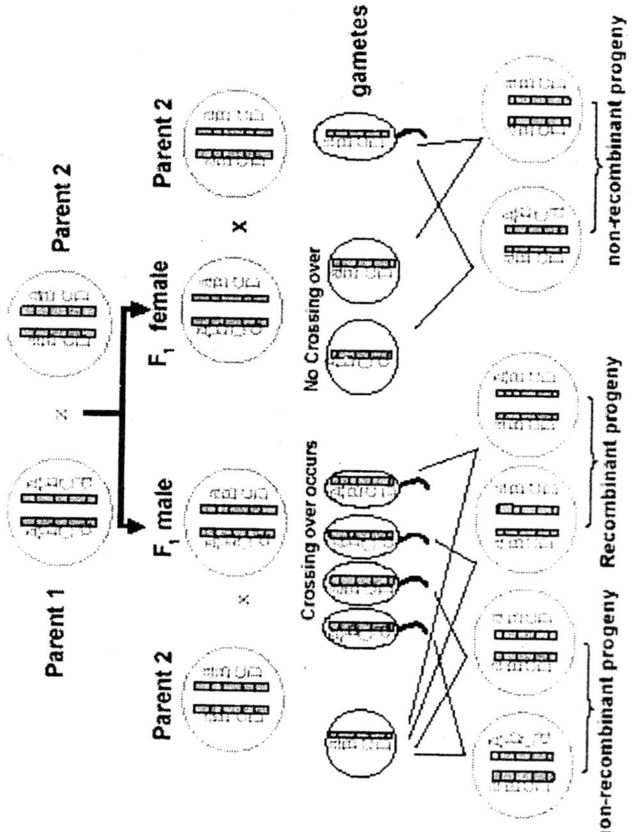

Figure 4. Biphasic nature of genetic linkage in silkworm

used by Promboon *et al.* The FISSR markers are particularly advantageous for high throughput genotyping in mapping of genomes. For constructing high-density linkage maps, a large number of F_2 individuals need to be analysed using a number of markers.

The template requirement in the case of other conventional PCR methods is relatively high. The FISSR assay provides a large number of DNA markers per primer and allows detection of markers with as little as 2–5 ng of template DNA, on an automated sequencer obviating the necessity for using radioactive isotopes.

Backcross populations have been used for mapping by three other groups. Nagarajahas used a polyvoltine strain, Pure Mysore, whose larval skin is normal white (+/+), and a bivoltine strain (*od/od*), whose larval skin is translucent, controlled by a recessive gene, *od*, located on the Z chromosome. Fifty five progeny were used for mapping. Tan *et al.*have used a similar back cross with 47 offspring. Similarly, Hara and coworkers have used a backcross strategy to make initial linkage assignments of codominant markers using a small number of segregants (15 offspring), to be followed up with a much larger number of progeny (on the order of 100) to develop a recombination map.

The use of backcross populations for mapping is well established in *Bombyx* genetics and recently in agricultural pests. It is advantageous since it avoids complications arising from the use of dominant markers and facilitates a biphasic approach to study linkage due to the occurrence of achiasmatic oogenesis (Figure 4). This approach was elegantly demonstrated recently by Heckel *et al.* in the mapping of a major gene conferring resistance to *Bt* toxin in the diamondback moth, *Plutella xylostella*, and in the tobacco budworm, *Heliothis virescens*, using AFLP markers. In these studies two backcross populations were created, one by crossing an F_1 female (derived from the initial cross between a susceptible strain and a resistant strain) with its susceptible parent and a second backcross population by crossing an F_1 male with the susceptible female parent.

From the first, population information on the linkage group contributing to the trait for resistance was derived. This is possible because in the absence of crossing over in the F_1 females, all of the genes present on the linkage group harbouring the gene for *Bt* toxin resistance would show absolute linkage. The second population was used to determine the location of the trait of interest in the corresponding linkage group because crossing over during gametogenesis in the F_1 males results in recombination of markers and thus allows one to determine map distance between markers. A similar strategy based on the molecular map could be used to track genes of economic interest in silkworm.

Quantitative Genetics

The concerted action of two or more individual genes as well as their interactions with each other and with the environment determine many important phenotypic characteristics. Such characteristics are referred to as polygenic or quantitative traits, and the individual gene locations as quantitative trait loci (QTLs). In silkworm almost all of the important economic traits such as fecundity, silk cocoon weight, silk fibre length, silk fibre thickness, and resistance to baculovirus are quantitative in nature. Very little is known about the number of genes involved, their chromosomal locations, or their gene products.

The development of high density genetic linkage maps that are based on DNA marker loci provides a useful tool for the resolution of these complex traits into their individual genetic components. Briefly, individuals from genetic crosses such as F_2 or backcross populations, or recombinant inbred lines or near isogenic lines obtained by repeated crossing of strains with contrasting phenotypes, are evaluated for the phenotype of interest and for their genotype classes at DNA-based marker loci at regular 10- to 20-cM intervals across the genome. The differences between the mean phenotype and marker genotype classes are then used to determine the genome positions and the effects of QTLs that contribute to the phenotype.

The DNA markers that are closely linked to the traits of interest in silkworm could be used to select the trait in question in a segregating population of an F_2 cross or using bulked segregant analysis, or with recombinant inbred lines or near isogenic lines produced in silkworm improvement programmes. Of course, the effectiveness of such DNA marker-assisted selection (MAS) will depend on the accuracy of the phenotypic classification of the trait of interest and the degree of linkage between a marker(s) and traits of interest in silkworm.

Cross-breeding strategies have been extensively used as a means of harnessing heterosis in the silkworm. In fact hybrid silkworms have contributed significantly to the dramatic increase in silk production of the world. An enormous amount of breeding effort has been invested in the development of silkworm hybrids, and this has resulted in the release of many hybrid combinations in Japan, China, India and other countries. While considerable success has been achieved in silkworm hybrid programmes, experimental data pertaining to the genetic basis of heterosis in silkworm have remained scarce.

The recent advances in genome research have generated considerable interest in predicting hybrid performance using molecular markers in crop breeding programmes. Most of the studies have been reported in corn and rice. A large number of studies conducted in maize have produced variable results: high correlations between molecular marker distances of the parents and hybrid performance were detected in some studies, while low correlations were observed in others. In rice also, there are extensive studies on the relationship between molecular marker heterozygosity and hybrid performance. As in maize, results are variable. Selection of genetically pure and divergent parental strains is critical to the success of a hybridisation programme in silkworm. The use of DNA markers to relate the genetic homozygosity and genetic distance of the parental strains to the manifestation of hybrid vigour, and understanding the genetic basis of heterosis, will be quite useful to select suitable parental lines for hybridisation programmes. Recently, molecular markers have been genotyped on a set of 22 silkworm strains crossed in a full-diallel programme to elicit the relationship between within and between strain genetic variability and the basis of mid- and better-parent heterosis.

The results indicate a positive correlation between genetic distance and mid-parental heterosis; there is also evidence to show that parental homozygosity has a distinct influence on the degree of manifestation of heterosis. Most important, the studies enabled them to identify the silkworm inbred lines which produced highly heterotic hybrids. Considering that a large number of silkworm inbred lines are maintained in the silkworm breeding stations in India and other silk-producing countries, the available molecular markers should be used to identify the parental lines that are most suitable for production of heterotic hybrids. The silkworm strain-specific markers could also be used for authentication of varieties, protection of breeders' rights, and periodical monitoring of the purity of varieties.

Physical Mapping

Isolation and characterisation of genes of interest using map-based cloning methods and systematic sequencing of the *Bombyx* genome by the proposed International consortium requires a high resolution physical map. A clone-based physical map consists of a set of ordered, overlapping inserts of cloned genomic DNA. Such a map affords a unique resource for studying the structure and function of the genome. The map facilitates the molecular identification of genes of interest, and the clones in the map provide ready access to substrates for genome sequencing.

A clone-based map also opens up new opportunities for studies of genome evolution. Clone-based maps have been assembled for several model organisms and agriculturally important species. In most organisms, sets of overlapping clones covering uninterrupted stretches of the genome (contigs) are assembled by detecting overlaps by means of shared restriction fragments in fingerprints or shared sequence tagged sites (STSs).

The construction of comprehensive and stable DNA libraries cloned in appropriate large-scale vectors is essential to obtain high-resolution physical maps of complex eukaryotic genomes such as silkworm. That the accuracy and efficiency of physical mapping are in direct relation to the mean size of cloned fragments in genomic libraries encouraged the development of several large-scale vector systems in recent years.

Yeast artificial chromosomes (YAC) were the first widely used systems, but many problems associated with clone stability, chimerism, and low yield led to the development of another large-scale cloning system based on the *E. coli* F factor which has largely superceded YAC-based cloning. These vectors are referred to as Bacterial Artificial Chromosomes (BAC). BAC libraries for silkworm have been constructed in three laboratories from two *B. mori* reference strains, p50 and C108, using nuclear DNA from posterior silkglands p50. Altogether these libraries contain more than 30 genome equivalents, with average insert sizes ranging from 120–170 kb.

The construction of BAC contigs using shared STSs, chromosome walking, and the fingerprinting methods is underway in the National Institute of Agrobiological Sciences, Tsukuba, Japan. In early studies, the previously reported close linkage between the *Bombyx* homologs of the *invected* (*Bm in*) and *engrailed* (*Bm en*) genes was confirmed by construction of a BAC contig that contained both sequences.

Initial screening of BAC libraries has also revealed information about the chromosomal organisation of two genes, *sericin-1*, which codes for one of the major cocoon proteins, and the gene encoding the precursor protein of diapause hormone and pheromone biosynthesis-activating neuropeptide, which are located within a 22-kb interval and are divergently oriented. In addition to their use to orient and organise known genes in fine-structure mapping of silkworm chromosomes, BAC contigs will serve as primary tools for large-scale sequencing of the silkworm genome. Initial efforts are underway to target selected portions of the genome, including chromosomes 1 (Z), 2, and W, with long-term plans to carry out a complete genome sequence.

3

Silkworm Egg Production

Silk raising begins with tiny silkworm eggs. Once they arrive place them in an open container like a wooden bowl or cardboard box. Keep in a warm place, but not in direct sun. They prefer the air to be about 85 degrees, well-ventilated, and moist rather than dry. They'll hatch in about two weeks.

The eggs hatch into very tiny furry black worms. They begin to eat immediately, and eat continuously except during molting periods, where they pause to shed their old, outgrown skin. There are four molts, at about two, three, and four weeks of age. At first, their food needs are small. But in the last two weeks as caterpillars, they consume enormous amounts of leaves. It is crucial to have plenty of mulberry for them.

The amount they eat directly transforms into the amount of silk they produce. Thus, these last two weeks of their caterpillar life are the most labour intensive for you. Life cycle of the silk caterpillar.

9 months	Eggs in the fridge
2 weeks	Eggs out of fridge to hatching

Contd....

4 weeks	Hatchlings through last molt
10 days	Last molt till begin to spin
3 days	Spinning
3 weeks	As pupa inside cocoon
1 week	As moth, mate, lay eggs, die happy
9 months	Eggs, in fridge

Success of sericulture depends on quality silkworm eggs. Therefore, management of seed production, transportation and incubation play important role on overall return. To produce quality seed, it is very important to adopt scientific methods of egg production right from seed crop rearing to egg incubation.

The grainage rooms along with its appliances should be thoroughly disinfected prior to commencement of operation and kept ready to receive seed cocoons. A day before disinfection, the rooms and appliances are to be washed with 5% bleaching powder solution; the appliances are to be sun dried for 3-4 hours. A day after, the rooms and appliances should be properly disinfected with a mixture of 2% formalin, 0.5 to 1% lime and 0.5% detergent solutions. Room is to be disinfected at the rate of 1 litre per square metre floor area.

The seed cocoons are to be always loosely packed either in perforated plastic crates or bamboo baskets or gunny/cloth bags and transported to respective destinations during cooler hours of day. However, transportation of seed cocoons to be preferred in crates or baskets.

Before selection of seed cocoons, it is very important to know the disease freeness of a lot, melted, flimsy and good cocoons are also to be separated. The gut portion is taken out and subjected to microscopic examination. In case there is incidence of pebrine, the lot has to be rejected. Before the arrival of new lots, it is very much important and essential to disinfect the contaminated rooms.

Immediately after the receipt of seed cocoons, they are to be spread on trays in a single layer to facilitate good aeration. Sorting of seed cocoons like melt, uzi infested, flimsy and those which are not conforming to the characteristics of parent races are to be taken out and rejected. The healthy seed cocoons alone should be preserved in trays for further processing, there should be cross ventilation in the preservation room, 25 ± 1°C temperature, 75 ± 5% relative humidity, 16 hour light and 8 hour dark conditions to be maintained in the cocoon preservation rooms. Complete darkness to maintain on the previous day of emergence, to avoid irregular emergence of moths.

Moths Eclosion

This helps in determining the disease freeness of a batch and helps in minimising the loss to grainage.

Early Eclosion

A simple box made up of wood and plywood sheet with a glass door is used having dimension 90 x 75 x 60 cm. The bottom is fitted with asbestos sheet. A heating element is connected to the electric main through a thermostat (0 - 60°C). On the top of the box at the centre, a 15 cm diameter ventilator covered with wire mesh is provided. It is fitted with sliding top to regulate ventilation. At the bottom of the box and on the lower portion of side walls, small holes are drilled to facilitate aeration.

To measure the temperature, a thermometer is fixed from inside of glass door. Within the box, a provision has been made to place 4-5 plastic trays in two tiers for keeping seed cocoons inside. For early emergence of moths, 50-60 seed cocoons are taken from individual lots and placed into an artificial eclosion box. The temperature in the box is adjusted to 32-33°C with the help of thermostat. This accelerates the development of pupae and moth emerges early. The early emerged female moths of respective lots are taken and subjected for microscopic examination to know the disease freeness of lots.

Synchronisation of Emergence of Moths, Pairing, Depairing and Oviposition

Before the expected day of emergence of moths, the cocoon preservation rooms should be kept dark. In case of variation in development of male or female pupae, the development of male pupae can be arrested by preserving them at 5-7°C and 75 ± 5% relative humidity for 3-4 days. Only healthy and active moths are taken for pairing. After 1-2 hours of emergence, the male and female moths of respective combination are allowed for 3.5 to 4 hours of pairing. At the time of depairing, the male and female moths are to be moved side ways so that the moths are separated easily without causing injury to reproductive organs.

The mated female moths are taken in a separate container and induced for urination. Moths are placed on egg sheet and covered with cellules and kept in semi dark condition for oviposition. Under proper preservation (5-7°C), male moths can be used for second pairing by giving 1-2 hours rest. Throughout the process of pairing, depairing and oviposition, optimum temperature of 25 ±

1°C and relative humidity of 75 ± 5% should be maintained.

Examination of Moth

The pebrine disease in *Bombyx mori* is caused by a parasitic sporozoan, *Nosema bombycis*. The mother moth examination can be taken up either (i) individual moth or (ii) mass moth examination.

Sterilisation of Eggs

After ascertaining the disease freeness of layings, egg sheets are dipped in 2% formalin for 10-15 minutes. This helps in removal of pathogens adhering to the egg shell and further prevents secondary contamination. Washing of eggs in formalin solution helps in firm adherence of eggs to the sheet.

Incubation of Eggs

Incubation facilitates uniform development of embryo. In addition, it greatly influences the voltinism of the eggs in succeeding generation, larval growth and success of cocoon crop. Therefore, the eggs are subjected to ideal conditions of incubation. Optimum temperature of 25 ± 1°C, relative humidity of 75 ± 5%, 16 hours of light and 8 hours darkness is ideal. During the pinhead stage or before two days of hatching, the eggs are black boxed to aim at uniform development of embryo and hatching of larvae at a time on a single day.

Preparation of Loose Egg

Loose eggs are getting popular for obvious advantages such as

- — Standard/uniform egg number (irrespective of the race, season, zone)

— Increased egg recovery and

— Easy and better management.

Methods

Dissolve 40 to 50g of maida in one litre of water and boil it to make it a paste. Smear the paste uniformly in a thin film on the craft paper sheet of appropriate size and dry

Spread the starched egg sheet in the oviposition tray so as to cover the entire floor surface besides the four of the frame.

— Spread the female moths uniformly at the rate of 40 to 50 (bivoltine) or 50 to 60 (multivoltine) per sq foot in the trays for oviposition.

— Keep the trays inside oviposition stand and allow the moths to oviposit under dark conditions.

— Maintain optimum temperature of 25 ±1 °C and relative humidity of 75 ±5 % for better egg recovery, both in terms of quality and quantity.

Eggs collection

— Conduct mother moth examination as per the standard procedure

— Soak the disease free egg sheets in water for 20 to 30 minutes to dissolve the gums

— Make ready the egg washing tray/collection unit by connecting to a water source, tie a nylon bag at the outlet for collection of eggs.

— Spread the soaked egg sheet on the egg washing tray and release water.

— Gently dislodge the eggs with hand.

— The eggs easily get dislodge and are collected in to the nylon mesh bag.

Removal of gum

— It is essential to remove the gum that is coated on the eggs.

— Prepare 0.3 to 0.5 % bleaching powder solution and allow it to settle.

— Release the eggs into the bleaching powder solution and wash for 10-20 minutes.

— Subsequently wash thoroughly in running water.

Eggs drying

— Collect the eggs into nylon bags.

— Gently squeeze out the water.

— Release the eggs on the platform of the drying unit.

— Spread the eggs in thin layer.

— Switch on the fan.

Acid treatment

— Collect the eggs in to nylon bags and tie the bags well.

— Dip the bag in to hydrochloric acid for the stipulated period (1.075 specific gravity, 46.1°C, 5-6 minutes) and gently disturb the eggs for uniform acid stimulus.

— After the treatment, withdrew the bag from the acid and wash thoroughly in running water for 20-30 minutes

— Dry the eggs as described earlier.

Eggs winnowing

- — For elimination of lighter eggs, feed the eggs into the hopper of the winnowing unit and switch on the fan.
- — Discard the lighter eggs which are blown away.

Brushing

- — As per the probable date of hatching, transfer the eggs two days earlier when they are pin-head stage into a paper boat/brushing frame.
- — Black box the eggs
- — On the expected day of hatching, spread the paper boat, distribute the eggs in a thin layer using a feather and expose to light.
- — Dust bed disinfectant on the hatched larvae.
- — Sprinkle adequately, chopped mulberry leaves on the net.
- — After one or two, lift the top net and transfer into a tray.

Eggs Handling

Depending on the need, the eggs could be made to hatch between 10 to 365 days by adopting different methods of preservation.

Artificial hatching

The bivoltine silkworm eggs before entering into dispause are acid treated with specific concentration of Hydrochloric acid at different temperatures.

Hot acid treatment

Hydrochlorisation of bivoltine silkworm eggs is a

common method to prevent eggs from entering into dispause. There is a correlation between the temperature and specific gravity of hydrochloric acid. When the hydrochloric acid is heated specific gravity falls. Adjustment of specific gravity hydrochloric acid ranging from 1.071-1.076 at normal room temperature of 25°C can safely be used for hot acid treatment of silk worm eggs, though ideal specific gravity at 46°C is 1.0642. The specific gravity ranging from 1.0715-1.0764 at room temperature 25°C to 1.0728 at 35°C room temperature are fit for acid treatment.

Adjustment of proper specific gravity of the acid is essential before starting acid treatment. Acid adjusted for the above specific gravity is heated to 46°C for 5-6 minutes for acid treatment. Hot acid baths specially made for this purpose are to be used where the principle adopted is to heat the acid kept in the container in a water bath indirectly. Some old hydrochloric acid used earlier for acid treatment is mixed with fresh acid for better results as fresh acid does not give desired results. The egg sheets are to be dipped in 2% formalin for about 2-5 minutes to make the eggs adhere firmly the egg sheets during hot acid treatment and to disinfect the eggs against pebrine.

The eggs are dried and later dipped in hydrochloric acid of same specific gravity as that used in treatment for 10 seconds and later treated in hot acid bath. This helps in maintaining the concentration and specific gravity of the acid during treatment of silkworm eggs. The eggs are washed in running cold water immediately after acid treatment to remote the acid completely as traces of acid on eggs lead to death of the eggs. Litmus paper may be used to check the presence of acid. The standard duration of hot acid treatment for various silkworm races is given below:

Japanese races	5-6 minutes
Chinese races	4-5 minutes
European races	6-7 minutes

For treatment of hybrid eggs the duration of duration of dipping in acid is determined as the mid-value of the duration of the two parents. Care must be taken not to exceed the duration of acid treatment as the eggs get killed and damaged. A proper alarm timer is to be used during acid treatment to avoid damage.

Preservation

Bivoltine silkworm eggs are preserved for hibernation for use in the next spring season. The eggs prepared in spring are kept in room temperature of 25°C with humidity of 75% to 805 till October-November. The temperature should not go higher than 30°C during this period and not lower than 20°C. The eggs reach dispause stage during this period. They are allowed to be kept in natural temperature which gradually falls to 10°C by December. If they are left in such conditions for long, they start hatching irregularly leading to unseasonal hatching. It is necessary to subject the eggs to low temperature of 7.5°C to 5°C for 50 to 60 days in order to ensure that all the eggs are hatch on time for rearing later in spring.

The eggs are generally kept in refrigerator of cold storage at 5°C from December to February. The eggs are transferred to 2.5°C-0°C for 40-60 days by middle of February. This helps in checking the growth of embryo. The eggs are kept at 10°C to 15°C for 4-5 days at the end of March or beginning of April and released for incubation or further cold stored, at 2.5°C till release at a later stage. The duration of preservation scheduled may be adjusted for early spring rearing in February to March. The eggs are released from cold storage and kept at normal room temperature for 3 to 6 hours for checking of sudden exposure to high temperature.

Seed Production Infrastructures

— Cocoon stage room

— Moth pairing room

— Oviposition room

— Moth Examination room

— Acid Treatment room

— Egg preservation room

Required Equipments

— Refrigerator

— Incubator

— Wooden Racks

— Wooden trays

— Antwells

— Working Stand and Tables

— Exhaust fans

— Moth crushing Sets

— Microscopes

— Round Bamboo Trays

— Acid treatment baths

— Thermometers/Hygrometers

— Seed production registers

— Specific gravity meter
— Glass-slides with cover
— Glass disc
— Glass pipette
— Measuring jar
— Water filter
— Egg sheets
— Cellules
— Wash basin

Required Chemical

— Formaldehyde
— Hydrochloric acid
— Potash
— Potassium Hydroxide
— Lime
— Bleaching powder
— Xylol

Rearing

Planning of the rearing amounts to the completion of work by 50%. It helps in synchronising mulberry growth, silkworm rearing in addition to augmenting productivity and material arrangement. The size and time of rearing mainly depends on the leaf availability in quantity and maturity. However, the rearing facilities to rear a unit quantity of eggs should be considered. The following points and parameters are to be considered for annual planning.

— Cropwise leaf production per unit area
— Rearing period
— Number of crops per year
— Rearing facilities
— Labour arrangement

Leaf Yield

The yield of leaf fluctuates depending upon the season. This aspect should be considered for estimating the quality of the layings to be brushed. 12-15 kg of leaves are required to rear 1dfl of bivoltine hybrid.

'Division of Plots

Single Plot System: This arrangement gives 5 to 6 crops per year. It requires one rearing house with leaf storage-cum-chawki rearing room.

Two Plots System: In this arrangement, rearing facility can be exploited two times and annually double the quantity of layings can be reared. But the important factor in this system is that a separate chawki rearing building is a must.

Mulberry Garden Management

Proper mulberry garden management has to be ensured so as to provide sufficient between two rearings (a minimum of ten days gap should be provided).

Rearing House

The rearing house is designed based on the brushing capacity and the method of rearing. In general, 2 sq.ft/dlf floor-area is required for tray rearing and 3 sq.ft/dfl in case of shoot feeding method. The same room can be used for chawki rearing and subsequently for leaf preservation in case of a single plot system. But in the case of 2 plot system, a separate chawki room away from the late age rearing house is essential. Each rearing house must have:

— A main rearing hall

— Ante room

— Leaf preservation room

Rearing houses are to be built in such a way to create a congenial atmosphere for the growth of the silkworm at the minimum operational cost. They are to be equipped with facilities to create required environment such as temperature and relative humidity. Raising evergreen trees, even mulberry trees around the rearing house help by maintaining a better rearing environment.

Rearing houses are to be designed for effective disinfection, like smooth surface of walls, use of lime white wash, facility for making the building air tight, and less reactive to disinfectants and washable smooth floor. The ideal location for leaf chamber is the north-east corner of the rearing house which is the coldest place of the building. It should not be well ventilate as it needs high moisture. The place must be cool, humid and dark.

Ante room

Ante room acts as barrier for entry of uzi fly, which should have minimum width and breadth of stretched arm (span) length of about 8 feet. The doors fitted with self-closing facilities are advantageous.

Chawki Rearing House

A separate chawki rearing rearing building with facilities

to raise temperature and humidity is a must for 2 plot rearing system. This may be located away from late rearing house and least accessible for the working group. The size of chawki rearing house is decided based on the quantity of eggs brushed. For a mulberry garden of 1 arce, the chawki room of 10'-14' is ideal. The size of the chawki rearing building should not be big. It is ideal to have false roofing with smooth surface material.

Preparation of rearing house

The rearing house should be kept ready in advance by disinfecting effectively using recommended procedure, at least 4 days in advance. Care must be taken to avoid accumulation of disinfecting chemical during rearing followed by preconditioning of the rearing house by arranging required numbers of rearing stands and trays and also adjusting rearing temperature and humidity one day in advance.

Rearing Appliances

Following are the appliances required for rearing silkworms for 1 arce of mulberry garden (300 dfls):

Non recurring (General)	Quantity
1. Disinfection mask and protective gum shoes	1 set
2. Sprayer (power sprayer, multipurpose pump or rocking sprayer for disinfection)	1 set
3. Room heater with thermostart 2 KV blowing type or charcoal stove	3
4. Water air cooler (Heavy duty)	3
5. Kerosene blow lamp	1
6. Wet and dry bulb thermometer	3
7. Forceps 6" long	3
8. Plastic sieve	1

	Non recurring (Specific)	Quantity
1.	Rearing building Chawki rearing room, leaf preservation-cum-preparation room, ante room Late age rearing hall, ante room, leaf preservation room	1
2.	Egg transportation box	1
3.	Egg incubation chamber	1
4.	Loose egg incubation frame	6
5.	Black box	1
6.	Chawki rearing trays (wooden 3'x4'x2.5')	24
7.	Chawki rearing bottom stand	2
8.	Rearing stands to accommodate 12 trays in arow	2
9.	Feeding stands	1
10.	Ant wells	72
11.	Leaf chopping board	1
12.	Leaf chopping knife	1
13.	Leaf mat (4' x 6' rexin)	2
14.	Leaf chamber	1
15.	Bed cleaning nets (3'x4', mesh size ¼ inch x ¼ inch)	48
16.	Earth pot	1
17.	Litter basket	2
18.	Late age rearing trays (bamboo 3.5′ dia) (for feeding method only)	120 for CB 140 for BV
19.	Rearing stands	12 for CB 14 for BV
20.	Feeding stands	3
21.	Shoot rearing rack 5'x35'x3 tiers (For shoot feeding method only)	2
22.	Mountages (bamboo chandrike)	120
23.	Plastic basin (big 18" dia and small 9" dia)	2
24.	Plastic bucket with lid (big 20 ltr cap.).	2
25.	Plastic mugs	2
26.	Plastic boxes with different colours	2
27.	Foam pads	50

27.	Foam pads	50
28.	Foot rugs	2
29.	Leaf chamber for late age	2
30.	Leaf basket (Bamboo)/(Plastic)	8/3
31.	Cleaning nets (4'x4')	280
32.	Cora cloth	8
33.	Gunnny cloth	10

	Recurring	Quantity
1.	Parafin paper (20 m/crop)	20
2.	Old new papers	10 kg
3.	Formalin	15 kg
4.	Bleaching powder	7 kg
5.	lime	25 kg
6.	Bed disinfectant	12 kg
7.	Power and fuel	
8.	Man power (Up to chawki rearing including leaf picking and after the IV instar only for rearing)	

	RearingLeaf	Rearing	Shoot
a.	Cleaning	3 men x1 day	3x1
b.	I age rearing	4x1	4x1
c.	II age rearing	2x4	2x4
d.	III age rearing	2x4	2x4
e.	IV age rearing	4x4.5	2x4.5
f.	Bed cleaning	-	2x1
g.	V age	6x8	3x8
h.	Harvesting	12	12
i.	Leaf harvest	54	
j.	Marketing	2	2

Transportation of Eggs

Silkworm eggs are to be handled with special care during transportation without desiccation. Usually, farmers use cloth bags, paper bags, etc to carry silkworm eggs. In these bags the eggs are directly exposed to various adverse climatic conditions, which lead to poor hatching of the eggs and poor health of the larvae. In addition to the physical compression, there are chances of contamination of toxic chemicals emanated from chilly, tobacco, detergents, etc.

To overcome this problem, a few simpler and cheaper transportation bags or egg carrying boxes utilising locally available materials are fabricated having a better humidity and temperature maintaining abilities. Each box or bag has provision of supplementing the required humidity and thermo-insulation property to avoid temperature fluctuations and provide rigid structure to avoid compression.

The bag is made or rigid plastic mesh lined with thin foam pad. It is supported by a rigid metal frame where as, the egg carrying box is made of rigid card board lined inside with thermocol and a provision to increase humidity and ventilation. Each box/bag has a capacity of 300 dfls. The transportation bag is kept wet till the transportation is complete. In box, wet foam pad is kept at the bottom for this purpose. It is ideal to transport eggs on the 2nd and 3rd day of the oviposition during cooler hours.

Egg Incubation

Providing ideal conditions to the eggs during embryonic growth for the complete, healthy and robust growth and development can be referred to as incubation. The silk worms eggs are to be incubated properly up to the 9th day from the day of oviposition and after examination of healthy development of embryos, and to avoid irregular hatching. Embryos grow robust using most of its reserve food under <25°C and >80% RH conditions. On the 9th day the egg sheets are re-arranged. About 25 dfls after disinfection are kept in the chawki-tray in single layer in incubator to enable brushing and proper spacing for young worms later.

The egg sheets are covered with sheet of black paper or paraffin paper. On the due date of hatching when a few larvae have started hatching, black paper or paraffin is removed and the eggs are exposed to dim light for about an hour. This facilitates uniform hatching of larvae. Care is to be taken not to over-crowd the eggs. Hygiene condition ensures prevention of disease contamination. Incubation should be thoroughly cleaned, washed and disinfected before incubation.

Low Cost Incubation Chamber (LCIC)

In order to maintain optimum temperature and humidity during embryonic developmental stage of eggs a simple electrically independent and easy to handle Low Cost Incubation Chamber (LCIC) is developed which is efficient in providing conducive conditions for the growth of silk worm embryo during incubation of silkworm eggs. LCIC, mainly consists of two chambers (outer and inner), the outer chamber of 21 inches height and opening at one end is made up of clay, of diameter of the top opening and 15 inches at the bottom closed end.

The wall of the outer chamber is provided with 5mm diameter holes at the top half-portion of the chamber, in rows in longitudinal pattern. The inner chamber is 18 inches in height and also made up of soil. The top end opening is having diameter of 15 inches that taper to 12 inches to the closed bottom end. This inner chamber is also provided with rows of holes of 5mm similarly. Clean and formalin disinfected sand bed of 1 inch depth is kept at the bottom of the outer and inner chambers. Sand in the chambers is kept moist sufficiently conditioned at least 12 hours before keeping eggs for incubation.

The egg sheets are aligned vertically with the help thin bamboo strip (with enough space for air circulation between sheets) and cover the mouth of both the chambers with loosely knit gunny cloth and preserve the eggs till black boxing. LCIC provides ideal temperature and RH for incubation of silkworms eggs. LCIC units can be prepared easily, as the material used for fabrication are locally available and cheap. The system is found to be very effective as it ensures optimum condition for the growth/development of silkworm eggs thereby hatching is optimised and higher is ensured.

Loose egg incubation

After transporting the loose eggs during cooler hours of the day, eggs should be incubated in incubation frames, which are made up of 2cm. Thick, light, wooden strips with two frames (outer and inner). The outer frame is 36x24cm and the bottom of the frame is lined with a thick black cloth. Inner frame is of 32x20 cm and it fits exactly inside the outer frame. One such frame can accommodate 1 box (20000 eggs or 50 dfls). Black boxing can be done in the incubation frame it self. Care should be taken to spread the eggs uniformly.

Black Boxing

Light has propound effect on hatching. Preserving the developed eggs in total darkness for a day or two before hatching is called black boxing. This helps in uniform hatching on a single day. During the black boxing, those embryos in advanced stage of development stop further development and this facilitates other to catch up.

Any device that provides complete darkness can be used for this purpose. Simple black sheet of paper (thick drawing sheets quality) or cover, which gives total darkness, is ideal. Eggs are grouped (25 to 50 dfls) in a tissue paper and covered on the day of pin head stage. They are transferred to black boxes. Such black boxes are placed under required humidity and temperature conditions. These eggs are exposed to light between 6 to 7 am.

Brushing

Transfer of the newly hatched larvae from eggs on the wax paper or rearing bed is called brushing. In 4'x3' tray, 50 dfls can be brushed and reared up to I moult and 25 dfls up to II moult. The egg sheets are distributed in required numbers per tray before hatching. Fresh, tender and succulent mulberry leaves (with 80-85 % moisture) from a well maintained garden are collected and chopped to a convenient size of 0.5 to 1.0 sq. cm. This is sprinkled over the hatched larvae on the egg card.

After 30 minutes, when all the worms have crawled on to the leaf, the leaves along with the worms are transferred on to the wax paper in the rearing tray with the help of a soft feather and arranged in the form of a thin bed of required size. Clean wet foam pads are kept around the rearing bed covered with another paraffin paper to maintain the required humidity. Pile up trays one above the in the form of a box. This arrangement conserves leaf moisture to a longer period for better feeding conditions, leading to vigorous and healthy growth of larvae.

In case of brushing loose eggs, inner incubation frame should be removed. Two layers of cotton/nylon net (mosquito net) is spread over the loose eggs and exposed to light. About 2 to 3 hours later, hairy small worms crawl out through the net. Then the chopped leaves are spread over the net. Cover the incubation frame with paraffin paper and keep undisturbed for an hour. Then the frame worms along with the top net is lifted from the incubation frame and brushed in a chawki tray. For the chawki rearing, blue polythene sheets can be used in place of paraffin paper.

The polythene sheet should be 400 gauge. During high humid conditions, wet foam pad and bottom sheet can be dispensed with. About 3 inch gap should be maintained between covering sheets and rearing bed. Finely chopped leaf bits can be used for brushing. This will help in controlling the initial thickness of bed effectively.

Handling of Leaf

Well grown fresh and succulent leaves (80-85% leaf moisture) raised under recommended package of practices, should be used for chawki stage. The leaf harvesting must be carried out during the cooler hours of the day and should be transported in clean, wet gunny bags or baskets to the rearing house. Store the leaf in humid place. During summer, the efficiency of leaf preservation is increased by sprinkling water on the leaves and then covering with wet gunny cloth

For young silkworm, the first two to three full blown leaves below the glossy leaf, which are tender, succulent and dark green in colour are suitable. As the worms grow, medium leaf should be used for feeding the worms. The nutritional conditions of chawki leaves can be maintained when they are preserved under low temperature, high humid and dark conditions. This can be effectively done by preserving them in earthen pots buried in wet sand.

Chawki Rearing

Rearing of young age silkworm up to III moult is called chawki rearing. Rearing of young age silkworms under ideal environmental conditions, feeding succulent, nutritious, tender leaves will make the larvae grow robust and make more stress tolerant during the advanced age of development.

Temperature

The silkworm body temperature will be little higher than the atmospheric temperature by 1? the ideal temperature for healthy growth of silkworm varies according to its developmental stage and variety. Increase of temperature increases the growth of silkworm. However, very high temperature imbalances the metabolic activity and the silkworm, thus becomes unhealthy. High temperature also affects the quality of leaf in the rearing bed. In case of lower temperature, the larval duration is extended due to low metabolic activities.

Finally, the silkworm may end up susceptible to diseases. Therefore, the optimum temperature for the chawki worms is 27-28 ? for I instar and 26-27 ? for II and III instars. Ideal temperature can be maintained by adopting the following methods. Low temperature can be increased by using electric heater preferably connected with thermostat or by burning charcoal with out emitting smoke. High temperature can be brought down by keeping the premise cool through shade, sprinkling, providing false ceiling or covering roof with dried plant material.

Humidity

Atmospheric humidity influences the silkworm growth mainly through mulberry leaf in addition to its direct effects on the silkworm. In case of low, the rate of multiplication of pathogens is slow and though hygienic conditions are better, the leaf withers fast, rendering it unfit for the silkworm to eat. As a result, growth of the larvae slow down causing inanition (weakness) with difficulty in moulting. On the other hand, high humidity keeps the quality of the leaf better, the larvae grow fast, becomes fat and weak in resistance to external conditions.

At high humidity conditions, pathogens multiply at higher rate,. Hence an optimum humidity during rearing seasons has to be maintained for healthy growth of the larvae and superior quality of the cocoons the optimum humidity is 85-90% for I and II instars. The humidity can be attained easily by using the following techniques:

— Use paraffin paper or blue polythene sheet as bottom and covering sheets in the rearing tray to conserve humidity.

— Keep clean, pathogen free and wet foam pads around the rearing bed.

Light

Light has varied effect on the silkworm behaviour and growth. Silkworm prefer dim light (20-30 lux). The intensity of this light should be uniform throughout the rearing building. The dim light during the day and darkness during the night is more congenial for healthy growth of the larvae.

Aeration

The rearing room is polluted by carbon dioxide, carbon monoxide, sulphur dioxide and ammonia due metabolic activity of the silkworm, mulberry leaves, workers and fermentation process in the bed. These unwanted gases have to be replaced by fresh air containing more oxygen for healthy growth of the silkworms. Keeping in view, stagnation of air should be avoided by proper ventilation.

Feeding and spacing

Chawki worms are to be fed 3-4 times a day with tender, succulent leaves. The quality of leaf to be fed each time should not be more than required; otherwise it leads to wastage of leaves. Before feeding, the bed have to be spread and dried to facilitate all the larvae to crawl on to the surface, which helps them to feed on fresh leaf. Further, it helps in maintaining better bed hygiene and micro-climate of the bed. As the larvae grow in size, they require more bed are.

Over crowding of the silkworms lead to competition of the available food space, which weakens the worms to a great extent. Regulate the spacing and feeding quality for healthy growth of the silkworms. Everyday after bed cleaning, optimal bed area has to be provided. Care should be taken for uniform distribution of larvae in the bed. It is important to remove paraffin paper cover and foam pads every morning prior to feeding. Chopsticks/ forceps should be used while handling the chawki silkworms.

Bed cleaning

As the larvae grow the unconsumed leaves and litter in the bed increases and may cause the multiplication of pathogens and release of obnoxious gases. Hence periodic cleaning of the larval bed is necessary. This is carried out using cleaning net with suitable mesh size.

Moulting and moulting care

Under optimum rearing conditions, the worms take 3 to

3½ days to settle for I moult and 2 to 2½ days to settle for the II moult. Moulting duration is 18 to 24 hours. When the worms start settling for the moult, the top paraffin paper and wet foam pads are removed. Feeding is also reduced during this period. When all the worms settle for moult, then slaked lime should be dusted @ 4-5 g/sq. ft. During moulting, the bed thickness should be reduced as far as possible. Adequate spacing allows the left-over leaf and bed for quick drying.

The above steps enable the moulting worms to settle uniformly. Increase in the bed humidity is harmful for moulting. Feeding is resumed half an hour later dusting bed disinfectants when more than 95% of the worms are out of the moult. Tender leaf is given for the first one or two feedings after the moult.

Disease management

Like any other animal, silkworm is also affected by various diseases caused by virus, bacteria and fungi. These diseases can be effectively prevented/controlled by killing the pathogens in and around the rearing house through disinfection before the commencement of rearing and by maintaining hygienic conditions and regular use of disinfectants during rearing. Chawki worms in general are more resistant to diseases than late age worms. However, the common diseases during chawki are fungal (muscardine and aspergillosis).

Diseases are more common when rearing bed is thick, bed humidity is high, worms kept in crowded conditions, less nutrition leaf, etc. As a control measures, the bed disinfectant has to be used, as recommended. In the event of outbreak of diseases, all the suspected/diseased larvae have to be removed carefully to avoid further contamination and placed in 2% formalin or lime solution before burning or buying. After this, the rearing bed is disinfected after every moult.

Maintenance of ideal temperature, humidity, aeration, spacing and feeding the larvae with nutritionally rich leaf improves hygiene and helps in prevention/controlling the diseases effectively. Periodic dusting of the mixture of bleaching powder and lime powder to the surroundings, floor and path leading to the chawki building helps in preventing the diseases.

New techniques in Chawki rearing

Various new technologies have been developed to simplify rearing and increase productivity and decrease the cost of production. Use of paraffin paper, foam pad and box rearing are some of the appropriate technologies incorporated for the chawki rearing. Some more new technologies have also been developed during recent years as detailed below.

Isolation Chamber: Successful chawki rearing needs ideal environmental conditions, good quality leaf and appropriate technologies. In the present situation, it is very difficult for the farmers to provide such ideal environmental conditions as most of them are conducting rearing in part of the dwelling house. Therefore, "Isolation chamber" has been developed top provide better environmental conditions and hygiene at the farmers level with low input cost, The isolation chamber can easily satisfy the average farmer as it can easily be accommodated inside his dwelling house. Isolation chamber can be made either with wood or brick masonry. Some advantages are:

— Uniform temperature and humidity can be effectively maintained with minimum power.

— Moisture loss in the leaf is minimised.

— Larval growth and moulting behaviour is uniform

— Chances of contamination with dust and pathogens minimised

— Larval duration is reduced

— Larval weight increases.

— ERR improves and enhances the crop stability.

Blue Polythene Sheet: To maintain optimum humidity conditions use of paraffin paper, wet foam pads and box rearing are popular in tropical conditions. Because of the natural wear and tear, high cost and non-availability of paraffin paper, the idea of using of blue polythene sheet came. It is used in silkworm beds in place of paraffin paper for bottom and cover, It is much cheaper and durable than paraffin paper. Some advantages are:

— Humidity for chawki rearing can be maintained in the bed.

— Driage of the leaf is reduced in the bed.

— Yield and cocoon characters are improved.

— Polythene sheets are easily available.

— Polythene sheets can be washed and disinfected.

— Polythene sheets can be used for more crops.

— It reduces the input cost.

Transportation of Chawki worms

The transportation of chawki silkworms reared in the chawki rearing centre is inevitable. It is always better if they are transported in the following conditions:

— Stage: Under II moult or two feeds after the II moult.

— Packing: Covered in wax paper.

— Time: During cool hours particularly in the evening.

— Distance: Shorter the better.

— As soon it reaches the rearing house, spread and keep them in the rearing stand.

Late Age Silkworm Rearing

The rearing of grown up silkworms, i.e. from 4th age of spinning is termed as late age rearing and is usually completed in 14-16 days. Silkworms attain sufficient growth during this stage, representing more than 94% of the leaf consumption, 133 times increase in body size, 125 times increase in body weight, and nearly 1000 times increase in silk gland weight. Moreover, the late age worms are more sensitive to high temperature and humidity conditions and diseases. Maximum growth and survival, which directly influence cocoon crop yield and this can be achieved only through scientific and skillful rearing.

Rearing house

A separate rearing house with adequate rearing space, sufficient ventilation and light is essential. This will enable effective disinfection and also to maintain required humidity, temperature and other hygienic conditions which helps to get good cocoon crop. The rearing building should be constructed in the outskirt of the village or in the garden in a centrally located place for easy access and to prevent cross contamination. The size of the building should be as per the requirement, with the height of about 12 ft. This helps in effective maintenance of temperature and humidity. The building should be fly and rodent proof with smooth finished walls, floor and rounded corners. Each rearing house should have an independent rearing hall with leaf storing room and ante-room.

Quality of leaf

The success of silkworms rearing mainly depends upon the quality of leaf. The IV and v instar worms require less moisture content in the leaf than the chawki worms. But the leaves should be nutritious. The worms are to be fed with medium and coarse leaf in the late age. Harvesting of the leaf is done during cooler hours of the day. Leaf can be either harvested by individual leaf plucking or in the form of shoots from the garden, which is 50 to 60 days mature.

After harvest the leaf is transported to the rearing room, packed loosely in wet gunny bags or bamboo baskets covered by wet gunny cloth. The leaves should be preserved properly in leaf chamber, covered with wet gunny cloth. The leaf moisture could be retained by spraying water over the leaf at frequent interval during summer. This enables the leaf to remain fresh for longer time and results on good cocoon yield.

Do not feed the worms with withered leaf as it leads to reduction in consumption and growth of worms. In the case of shoot preservation, mulberry shoots should be preserved vertically and covered with clean wet gunny cloth in dark room with low temperature and high humidity.

Environmental conditions

Though temperature range of 24 to 26°C is ideal for rearing late age worms, the rearing cold be successfully conducted even at higher temperature by proper manipulation through effective cooling, regulation of humidity, frequency of feeding and aeration.

Bed spacing and feeding

Crowding of silkworm in the rearing trays and poor ventilation leads to the building up of toxic gases in the rearing room, which is harmful to the silkworms. Proper spacing and good aeration keeps the worms healthier. Crowding of worms in the rearing trays also leads to under-nourishment and poor growth resulting in low cocoon yield. Rearing trays should not be smeared with cow dung.

Care during moult

Silkworm takes 4 to 4½ days in IV age to settle for moult. It takes nearly 30-36 hours in IV moult. At the time of moulting, the worms require comparatively dry atmospheric conditions. To achieve this, keep the bed thin, spread the bed gently, dust lime powder over the bed, provide good spacing and increase aeration in the rearing house. This reduces the bed humidity and dries the left over leaves quickly.

As the worms show signs of moulting, the feeding for moult, should be reduced. When 90 to 95% of the worms settle for moult, feeding should stop. Resumption of feeding of only done after 95% of the worms have come out of the moult. Tender leaf is given at the time of resumption after applying disinfectant. Unsettled worms should be separated and rejected before dusting the lime. If the unsettled larvae are more and are healthier, keep them separately, give feeding and allow to settle for moult.

Bed cleaning

Rearing worms in tray system requires bed cleaning

everyday during instars. Cotton/nylon nets can be used for bed cleaning.

Crop protection

In the uzi fly infected area, silkworm rearing must be conducted in well protected nylon net enclosure or in fly proof rooms. Use of uzicide, uzitrap, uzi powder or any other recommended chemical reduces the uzi fly infection. Recommended bed disinfectants which can be used to prevent grasserie and muscardine diseases are applied half an hour before resuming feed after each moult and once again on the 4th day of V age.

Mounting and harvesting

At the end of 5th age i.e., on 6th or 7th day, larvae reduces feeding, shrink in size, body becomes translucent and change their colour to light yellow and start crawling in the bed with raised head and release wet faecal matter,. This is the indication of spinning larvae, and thereafter the feeding quantity should be reduced and aeration should be increased. Once the spinning larvae appear in the bed, pick them by hand and put them on mountages. Take away the big branches with leaf, if any, in the bed before picking spinning larvae.

When more of the worms are matured for spinning, they can be easily separated by shaking the branches and collecting the larvae enmass. Net can also be used for separating the spinning larvae. Mount the spinning larvae on bamboo mountages. Temperature of 24°C and humidity of 60-65 % and aeration are ideal for spinning. Therefore, take necessary care during spinning time to get quality cocoons. Harvest cocoons on the 5th or 6th day after spinning during summer seasons, respectively. For better cocoon quality and quantity, rotary mountages are ideal.

Feeding of Late Age Silkworms

Sericulture is highly labour intensive, contributing more than 50 5 of the cost in silkworm rearing, further, the labour requirement particularly of leaf harvest is also very high. Keeping in view, shoot feeding technology is recommended for late age silkworm to enhance margin of profit in sericulture.

The popular silkworm rearing, in general, is by harvesting individual leaf and feeding them to silk worms in trays which are kept 10 to 12 in number one above the other in a rearing stand up to a rearing stage. This traditional method of rearing involves high labour of leaf picking, feeding and for everyday bed cleaning. In the shoot feeding method, the last two stages of rearing is done, by giving whole mulberry shoot instead of individual leaves, and it is out in plat form type rearing rack. There will be one bed cleaning after 4th moult.

Rearing and appliances

A separate house preferably in the garden or nearer to the garden is convenient. The rearing house should be built on a raised platform with shady trees around to provide good aeration and cool surroundings. The rearing house should be rodent and uzi proof with an ante-chamber and a shoot storing-cum-chawki room. Approximately, 3 sq. ft. per dfl floor area is the requirement.

Transportation and marketing of cocoons

After harvesting, the cocoons should be cleaned by removing litter. Double cocoons and flimsy cocoons are separated out and transported in loosely filled gunny/ cloth bag during cooler hours of the day for marketing.

4

Characteristics of the Cocoon

Physical Characteristics

The silk glands of the *Bombyx mori* are structured like tubes consisting of a Posterior, Middle and Anterior section. The Posterior is long and thin. The Middle is short with a diameter measuring 3-4 mm. The Anterior is extremely thin, leading to the spinneret in the head of the larvae from which the silk is excreted. Fibroin is secreted in the Posterior and transferred by peristalsis to the Middle section, which acts as a reservoir. Here it is stored as a viscous aqueous solution until required for spinning.

The majority of the sericin is created within the walls of the Middle section. In fact, these two proteins are reserved side by side in the Middle section without mixing one into the other. The fibroin core is covered with a layer of sericin and the secretions from the two proteins join at the junctions where the sericin is fused into one layer. The Filipis glands discharge a liquid protein. To form its cocoon, the silkworm draws out the thread of liquid protein and internally adds layer after layer to complete this protective covering.

Colour

Colour is a characteristic particular to the species. It is the presence of pigments in the sericin layers, which cause the colour. This colour is not permanent and washes away with the sericin during the degumming process. There are diverse hues of colour including but limited to white, yellow, yellowish green and golden yellow.

Shape

Cocoon shape, as colour, is peculiar to the given species. At the same time shape can be affected by the execution of the mounting process, especially during the cocoon spinning stage. Generally, the Japanese species is peanut-shaped, the Chinese elliptical, European a longer elliptical and the polyvoltine species spindle-like in appearance. Hybrid cocoons assume a shape midway between the parents for example, the case of a longer ellipsoid or shallowly enclosed peanut form. The shape of cocoons assists in identifying the variety of species plus evaluating reelability.

Wrinkle

The deflossed cocoon has many wrinkles on its surface. Wrinkles are coarser on the outer layer than within the interior layer. The outline of the wrinkle is not uniform, but various according to species and breeding conditions. Spinning employs high temperature and low humidity settings, which render fine wrinkles or cotton-like textures of, cocoon layers. These provisions discourage the agglutination of the baves resulting from accelerated drying. It is recognised that coarse wrinkled cocoons reel poorly.

Size

Cocoon size or volume is a critical characteristic when evaluating raw materials. The size of the cocoon differs according to silkworm variety, rearing season and harvesting conditions. The number of cocoons per litre, ranging between 60 and 100 in bivoltine species calculates size. Multivoltine species measure considerably higher.

Weight

The most significant commercial feature of cocoons is weight. Cocoons are sold in the marketplace based on weight as this index signals the approximate quantity of raw silk that can be reeled. The whole weight of a single cocoon is influenced by silkworm species, rearing season and harvest conditions. Pure breeds range from 2.2 to 1.5 g, while hybrid breeds weight from 2.5 to 1.8 g.

Table 1. Daily loss in weight of fresh cocoons

Days after mounting	6	7	8	9	10	11	12	13
Days after pupation	2	3	4	5	6	7	8	9
Index of fresh cocoon weight	100	99.4	98.8	98.3	97.7	97.0	96.1	95.1

In nature, the weight of a fresh cocoon does not remain constant but instead continues diminishing until the pupae transforms into a mother and emerges from the cocoon. This weight occurs gradually as moisture evaporates from the body of the pupae and as fat is consumed during the metamorphosis process (Table 1).

Thickness

The thickness of the cocoon shell is not constant and changes according to its three sections. The central constricted part of the cocoon is the thickest segment, while the dimensions of the expanded portions of the head are 80 to 90 percent of the central constricted (Table 2). The weight of the silk shell is the most consequential factor as this measure forecasts raw silk yield. Shell weight differs in correspondence to varieties of silkworms.

Table 2. Variations of shell thickness in different parts

Varieties		*A*		*B*		C
Parts micron	*index*	*micron*	*index*	*micron*	*index*	
Head extreme end	336	47	302	43	228	40
expanded	594	83	596	85	480	84
Central constricted	712	100	700	100	572	100
Tail expanded	568	80	590	84	466	81
extreme end	382	54	333	48	348	43

Further, weight is also influenced by the type of technology used for rearing and mounting. In practice, uni and bivoltine species produce heavier shell weights than multivoltine species.

Hardness

Cocoon hardness correlates to shell texture and is affected by cocoon spinning conditions. For instance low humidity during the mounting period makes the cocoon layer soft, while high humidity makes it hard. The degree of hardness also influences air and water permeability of cocoons during boiling. A hard shell typically reduces reelability (during the cocoon reeling process), while a soft-shell may multiply raw silk defects. In short, moderate humidity is preferred for good quality cocoons.

Shell Percentage

As the entire cocoon including the pupa is sold as part of the raw material, it is essential to quantify the ratio of the weight of the silk shell versus the weight of the cocoon. This is calculated in the formula:

$$\frac{\text{Weight of the cocoon shell}}{\text{Weight of the whole cocoon}} \times 100$$

This value gives a satisfactory indication of the amount of raw silk that can be reeled from a given quantity of fresh cocoons under transaction. The calculation assists in estimating the raw silk yield of the cocoon and in deriving an appropriate price for the cocoons. The percentage will change based on the breed of the silkworms, rearing and mounting conditions. Percentage rates are altered based on the age of the cocoons as the pupa loses weight as metamorphosis continues. In newly evolved hybrids, recorded percentages are 19 to 25 percent, where male cocoons are higher than female cocoons.

Raw Silk Percentage

This index is the most important for the value of the cocoon as it has a direct impact on both the market price of cocoons and the production costs of raw silk. The normal range is 65 to 84 percent for the weight of the cocoon shell and 12 to 20 percent for the weight of the whole fresh cocoon.

Length of Filament

Equally important as the percentage of silk shell is measuring the length of the bave contained in the shell. The factor determines the workload, rate of production, evenness of the silk thread and the dynamometric properties of the output. The length of cocoon filament corresponds to the varieties of silkworms. Range of total length is from 600 to 1 500 m of which 80 percent is reelable while the remainder is removed as waste.

Reelability

Reelability is defined as the fitness of cocoons for economically feasible reeling. Industry practice measured the case with which the cocoon yields the bave in reeling. Poor reelability causes a variety of production problems such as halts in production due to filament breakage and high degrees of waste product. Reelability is greatly affected by careful action during cocoon spinning, drying, storage, pre-processing, reeling machine efficiency and operator skill.

$$\text{Reelability (\%)} = \frac{\text{Number of reeled cocoons}}{\text{number of ends feeding}} \times 100$$

Recent statistics show an average reelability of percent for good cocoon varieties. The measured range is from 40 to 80 percent with serious deviations depending on the type of cocoon. Note that stained cocoons generally have poor reelability.

Size of Filament

The measure denier expresses the size of silk thread. A denier is the weight of 450 m length of silk thread divided into 0.05 g units. The diameter of the bave is not constant throughout its length, instead changes according to its position in the bave shell.

At the coarsest section of cocoon filament from 200 to 300 meters, the denier increases. Once more these dimensions become finer and finer as the process approaches the inside layer (Figure 1). The average diameter of cocoon filament is 15 to 20 microns for the univoltine and bivoltine species.

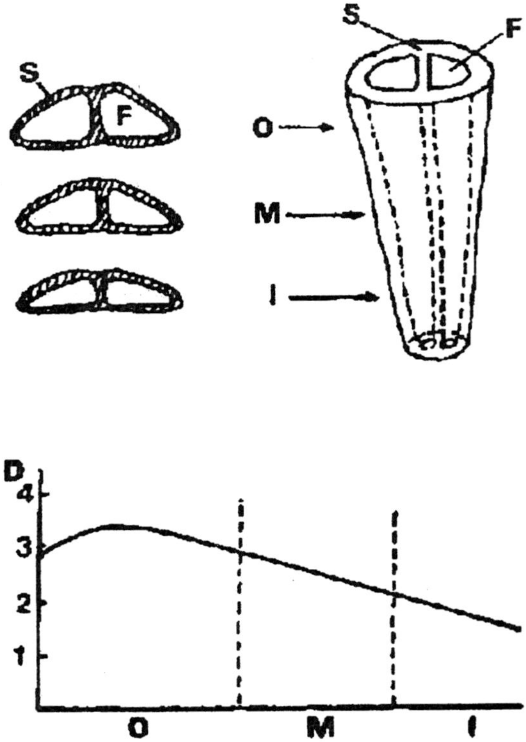

Figure 1. Size curvature of cocoon bave S: sericin; F: fibroin; O: outer layer; M: middle layer; I: inner layer; D: size (denier)

Defects

A series of minor defects may be found in cocoon filament such as loops, split-ends, fuzziness, nibs and hairiness (Figure 2). While these defects are observed among silkworm varieties, mounting conditions seem to contribute to their incidence. These filament defects directly affect raw silk quality. It is not recommended that silk varieties graded below 90 percent in the Neatness Chart be used.

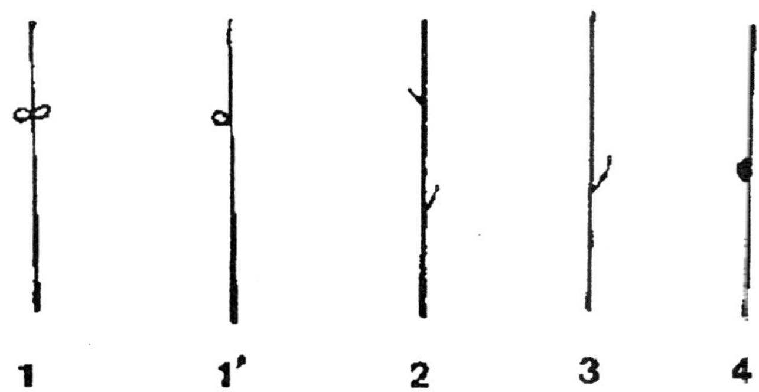

Figure 2. Neatness defects (1) (1') Loop (2) Hairiness (3) Split-ends (4) Nibs

Lousiness

Hair-like projections in the silk fibre are called Lousiness. Lousiness is more prevalent in baves produced by silkworms, which have been overfed in their fifth stage of rearing. Lousiness is found less in breeds of silkworms, which spin finer bave. Another factor promoting lousiness is mounting of over-mature larvae. This defect poses serious problems to silk fabric manufacturers, in particular those producers of smooth satin and necktie materials. When fabrics woven with these defects are dyed, it looks as if the fabric is covered with dust or is a paler shade than the rest. In fact, the protruding fibril is more transparent and has a lesser capacity to absorb dyes.

Composition

The composition of the whole cocoon is defined as the cocoon shell, pupa and cast off skin shown in Table 3. The pupa makes up the largest portion of its weight. Note that much of the cocoon content is water; therefore it is necessary to remove the water to improve the cocoon filament for reeling and to better preserve the cocoon over a long period.

The silk filament forming the cocoon shell is composed of two brins (proteins) named fibroin and covered by silk gum or sericin. The amount of sericin ranges from 19 to 28 percent according to the type of cocoon.

Table 3. Composition of the cocoon

Weight	*Fresh Cocoon*				*Dried Cocoon*			
	Race A		*Race B*		*Race A*		*Race B*	
	Actual number (g)	*Ratio (%)*	*Actual number (g)*	*Ratio (%)*	*Actual number (g)*	*Ratio (%)*	*Actual number (g)*	*Ratio (%)*
Cocoon	2.181	100.0	2.156	100.0	0.851	100.0	0.888	100.0
Cocoon shell	0.404	18.5	0.458	21.2	0.398	46.8	0.452	50.0
Pupa	1.765	80.9	1.684	78.1	0.441	51.8	0.422	47.5
Cast-off skin	0.012	0.6	0.014	0.7	0.012	1.4	0.014	1.6

The composition of the cocoon shell is given below:

Fibroin	72-81 percent
Sericin	19-28 percent
Fat and wax	0.8-1.0 percent
Colouring matter and ash	1.0-1.4 percent

Usually the sericin content of the cocoon shell is at the maximum level at the outside layer 1 becoming progressively lower at the middle layers 2 and 3 and the absolute minimum at the inside layer 4 (Table 4).

Table 4. Sericin content to different layers of cocoon shell

Cocoon layers	*Race A (%)*	*Race B (%)*	*Race C (%)*	*Race D (%)*
Outside 1	31.40	32.08	34.13	33.15
2	23.45	29.29	27.50	27.71
↓ 3	20.11	22.22	23.96	23.47
Inside 4	18.12	20.63	21.54	21.33

Properties of Silk

The silk of *Bombyx mori* is composed of the proteins fibroin and sericin, matter such as fats, wax, sand pigments plus minerals. Fibroin in the *Bombyx mori* comprises a high content of the amino acids glycine and alanine, 42.8 g and 32.4 g respectively as shown in Table 5. The key amino acids in sericin are serine (30.1 g), threonine (8.5 g), aspartic acid (16.8 g) and glutamic acid (10.1 g) (Table 5).

Table 5. Amino acid composition of Fibroin and Sericin

Amino acids	*Fibroin*	*Sericin*	*Amino acids*	*Fibroin*	*Sericin*
Glycine	42.8	8.8	Glutamic acid	1.7	10.1
Alanine	32.4	4.0	Serine	14.7	30.1
Leucine	0.7	0.9	Threonine	1.2	8.5
Isoleucine	0.9	0.6	Phenylalanine	1.2	0.6
Valine	3.0	3.1	Tyrosine	11.8	4.9
Arginine	0.9	4.2	Proline	0.6	0.5
Histidine	0.3	1.4	Methionine	0.2	0.1
Lysine	0.5	5.5	Tryptophan	0.5	0.5
Aspartic acid	1.9	16.8	Cystine	0.1	0.3

Sericin is a complex protein composed of three distinct components (I, II and III) of which sericin III is the interior layer directly adjacent to the fibroin core. The sericin I outer lay is the most soluble of the three constituents, while sericin III is difficult to dissolve. Viewed as a cross section, the brins have the appearance of equilateral triangles with rounded corners that face each other at their respective bases (Figure 3).

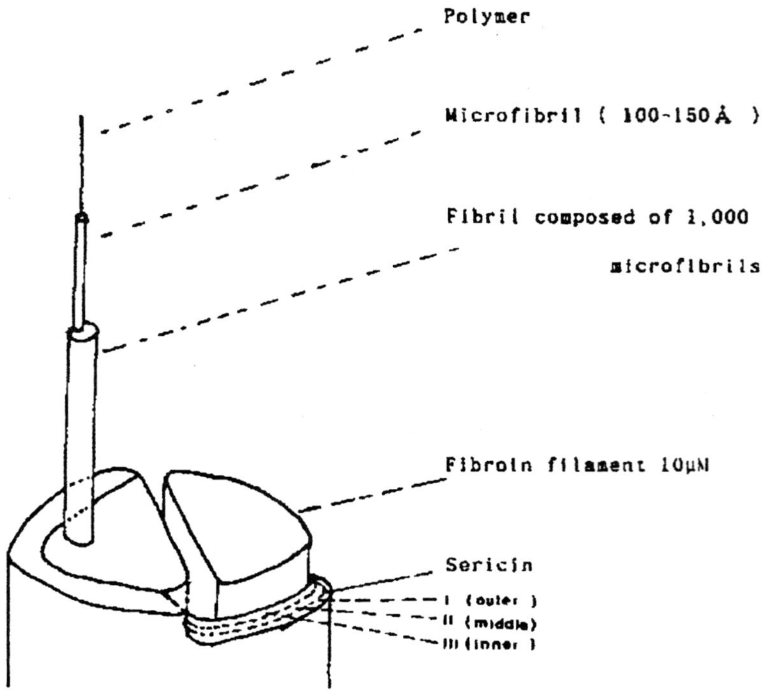

Figure 3. Texture of the silk thread

When the brin is crushed, it splinters into numerous minute fibrils revealing the actual structure of the brins. The thickness of each fibril is less than one micron and they are parallel to the axis of the fibre. A single fibril contains many microfibrils which, when examined with an electron microscope, have a diameter of approximately

100 Å per microfibril. Microfibrils contain micelles, which are separated into crystalline and amorphous segments.

Specific Gravity

The bave specific gravity on average of sericin and fibroin measures from 1.32 to 1.40. Generally, the specific gravity of sericin is slightly higher than that of fibroin (Table 6).

Table 6. Specific gravity and tensil strength of various fibres

Fibres	Specific gravity	Tenacity (g/denier)	Elongation (%)
Raw silk	1.32-1.40	2.6-4.8	18-23
Degummed silk	1.30-1.38	-	-
Wool 16	1.30-1.40	1.2-1.5	30-48
Cotton	1.52-1.60	3.2-4.8	7-11
Flax	1.50-1.58	4.8-6.0	2-4
Nylon	1.14-1.17	4.5-5.0	25-30

Tenacity and Elongation

Tenacity indicates the quantity of weight a given fibre can support before breaking. the typical tenacity of a bave is 3.6 to 4.8 g per denier (Table 6). Degummed silk has greater tenacity than raw silk. Elongation defines the length to which a fibre may be stretched before breaking. Raw silk has an elongation of 18 to 23 percent of its original length. Excess moisture increases the elongation of silk, but decreases its tenacity.

Hygroscopic Nature

$$\text{Denier(D)} = \frac{\text{Fibre weight (g)}}{\text{Fibre length (m)}} \times 9000$$

Moisture content and humidity are of critical importance to commercial silk production. Desorption measures a greater regain at a given relative humidity. For instance, given 65 percent RH, the adsorption regain value is 10 percent and the associated desorption value is 11.1 percent. Currently, 11 percent is the accepted moisture regain coefficient for silk; the mercantile weight of silk is derived based on this factor.

Influence of Light

Continuous exposure to light weakens silk faster than cotton or wool. Raw silk is more resistant to light than degummed silk. It is advised that silk drapery and upholstery fabrics be protected from direct exposure to the light.

Electrical Properties

Silk is a poor conductor of electricity and accumulates a static charge from friction. This trait can render it difficult to handle in the manufacturing process. This static charge can be dissipated by high humidity or by maintaining a R.H. of 65 percent at 25°C. Based on its insulating properties, silk is used extensively for covering wire in electrical equipment.

Action of Water

Silk is a highly absorbent fibre, which readily becomes impregnated with water. Water, however, does not permanently affect silk fibre. Silk strength decreases about 20 percent when wet and regains its original strength after drying. The fibre expands but does not dissolve when steeped in warm water. Note that the fibre will also absorb dissolved substances present in water. This is the reason that special attention is given to the quality of the water utilised for reeling, washing, dyeing or finishing.

Influence of Heat

If white silk is heated in an oven at 110°C for 15 minutes, it begins to turn yellow. At 170°C, silk disintegrates and at its burning points releases an empyreumatic odour.

Degradation by Acids, Alkalis

Treatment of silk fibres with acid or alkaline substances causes hydrolysis of the peptide linkages. The degree of hydrolysis is based on the pH factor, which is at minimum between 4 and 8. Degradation of the fibre is exhibited by loss of tensile strength or change in the viscosity of the solution.

Hydrolysis by acid is more extensive than alkali, and it has been postulated that acid hydrolysis occurs at linkages widely distributed along the protein chain, whereas in the early stages of the alkaline treatment, hydrolysis happens at the end of the chain. Hydrochloric acid readily dissolves fibroin especially when heated – and this is used mainly in studies of hydrolysis. Hot concentrated sulphuric acid, while rapidly dissolving and hyrolysing fibroin, also causes sulphation tyrosine.

Nitric acid readily decomposes fibroin, due to its powerful oxidising properties and concurrently causes nitration of the benzene nuclei. Organic acids have few effects at room temperature when diluted, but in a concentrated form fibroin may be dissolved, along with a certain amount of decomposition.

Proteolytic Enzymes

Proteolytic enzymes do not readily attack fibroin in fibrous form apparently because the protein chains in silk are densely packed without bulky side chains. Serious degradation may be caused by water or steam at 100°C.

Oxidation

Reports regarding the oxidation of proteins are rather meagre since the reactions are very complex. Oxidising agents may attack proteins in three possible points:

a) at the side chains,

b) at the N-terminal residues, and

c) at the peptide bonds of adjacent amino groups.

Hydrogen peroxide is absorbed by silk and is thought to form complexes with amino acid groups and peptide bonds. It has been demonstrated that hydrogen peroxide diminishes the tyrosine content and further that the peptide bonds are broken at the tryosine residues. Peracetic acid causes more rapid scission and produces more acid groups than peroxide.

Other Agents

Chlorine attacks fibroin more vigorously than does sodium hypochlorite. The oxidation is mainly at the tyrosine residues.

5

Cocoon Quality and Classification

Cocoon is a pupal casing made by moth caterpillars and other insect larvae.

Cocoon Quality

A series of natural circumstances will produce variations in cocoon quality. Some of the most noteworthy include:

— differences in cocoon quality in the same batch;

— differences in cocoons produced in the same location by different farmers who have reared the same species;

— seasonal influences. In Japan for example, cocoons produced in the spring and late autumn are higher in quality than those in early autumn and summer;

— environmental conditions affect cocoon reelability such as temperature and humidity;

— processing technique in reeling will impact reeling efficiency as well as raw silk quality;

— bivoltine cocoons are superior quality compared to multivoltine silkworm species traditional farmed in tropical zones.

Recent silkworm cultivation now develops cross-breeds of multivoltine with bivoltine silkworms as a strategy to improve overall cocoon quality.

Factors Influencing Cocoon Quality

Temperature and Humidity

Maintain temperatures at or near 25°C and relative humidity around 65 percent for silkworms to spin good quality cocoons with a high reelability.

Mounting Device

Although different mount practices are employed among producer countries, rotary mounting frames provide good ventilation. The result is improved reelability of cocoons.

Harvesting and Handling

Cocoons should be harvested only following complete pupation. In practice, the appropriate harvesting day would be the fifth day in tropical countries, and the seventh or eighth day in temperate countries, from the mounting date. If premature harvesting takes place, the silkworm will still be in its larval stage, weigh more, have fragile skin, and could likely be crushed, which would cause stains to the cocoon during handling and transportation.

Transportation

After proper harvesting and removal of diseased or damaged cocoons, the fresh cocoons are taken to the market. For short distances, the farmer carries the cocoons in bamboo baskets or jut bags on his head or by bicycle. If the distance is longer, cocoons are transported in a van or a bus. Caution should be exercised when loading fresh cocoons on to the van to ensure that containers are loosely packed in tiers to avoid damage. Vibration and shock during long trips can spoil fresh cocoons.

Cocoon quality is affected by steam produced while being transferred in a bag or basket. If there are defective cocoons fresh cocoon quality will be harmed. Table 1 describes the impact of transportation on fresh cocoons. While it is advisable to avoid carrying cocoons over long distances, there are steps, which preserve silk reelability. First use of P.V.C. containers with 15 kgs capacity is recommended.

Table 1. Effect of transportation containers of fresh cocoons on reeling results

Transportation containers	*Raw silk yield (%)*	*Cocoon reelability (%)*	*Reeling troubles per 10,000m (times)*	*Neatness (%)*	*Cleanness (%)*
Bamboo basket with cotton bag	18.50	71.1	2.17	93.0	95.0
P.V.C. container with sponge (shock absorber)	18.75	75.0	1.83	96.5	96.5

Shock absorbers, such as sponge can prevent damage over long distances. To minimise the risk of heat deterioration, shipping should take place only during the night or early morning. Ideally, the fresh cocoons should arrive at the stifling unit within two to three days after harvest.

Cocoons Classification

When cocoons are sold at the market, price is assessed on the basis of cocoon quality. This is judged by grading shell percent, filament length, reelability and the percentage of defective cocoons. If the percentage of defective cocoons is high, the price will be affected.

Defective Cocoons

Double Cocoons

A double cocoon is spun by two worms, producing a filament, which does not unwind smoothly and tangles easily. As these cannot be reeled along with normal cocoons, double cocoons are used for manufacture of a coarse, non-uniform, stubby yarn called "doupion". Double cocoons may be caused by crowded mounting conditions, high temperatures, high humidity and mutation of silk species.

Dead Cocoons

Dead cocoons are also known as melted cocoons. In this case, the pupa is dead and sticks to the inside shell of the cocoon causing a stain. Melted cocoons are called mutes because they do not make a sound when shaken. These cocoons are difficult to process and will result in silk, which is dull in colour.

Outside Stained Cocoons

These are recognised by a rusty colour spot on the cocoon shell caused by absorption of intestinal fluid/urine of the mature worm formed during mounting. Reelability is very poor in this case.

Printed Cocoons

This defect may happen due to improper mounting frames; these are also called scaffold pressed cocoons.

Malformed Cocoons

These are abnormally shaped cocoons, which may arise from species variation. This defect may be due to racial characteristics and breeding with mulberry leaves stained with agrochemicals.

Flimsy Cocoons

Here, the shell is loosely spun in layers and has a low silk content. These cocoons are easily overcooked and produce waste.

Thin-end Cocoons

One or both ends of the cocoon are very thin and risk bursting when processed. The cause of this defect may be attributed to species characteristics or improper temperature and humidity during rearing and mounting.

Pierced Cocoons

This happens when a moth has emerged, been eaten by beetles or in the case of the emergence of a parasite. Pierced cocoons are unfit for reeling and can be used only for hand spinning or as raw material of machine spun silk yarn.

Testing and Grading

In sericulturally advanced countries, cocoons are subjected to systematic testing and grading before sale. Prices are based on the quality of the cocoons. But in developing countries, there is no system for cocoon testing. Cocoons are sold on visual inspection and personal experience is relied upon in marketing of cocoons. No laws exist or compulsory testing and trading of cocoons. The result is that cocoons are simply auctioned or in certain instances, even sold at a price fixed by the Regional Departments of Sericulture. There is no direct correlation between price and quality of cocoons.

Major sericultural countries through the International Sericultural Commission (ISC) have studied an International Cocoon Classification System. However, diversion of silkworm species, techniques of breeding, silk reeling and other factors lead to the production of non-uniform cocoons. These parameters have challenged the effort to unify equipment and testing methods (Table 2).

Cocoon testing and grading may be accomplished with a compact automatic reeling machine as well as a multi-ends reeling machine, which is typical equipment in major sericultural countries. Testing methods used by major sericulture countries are displayed in Table 2. The quantity of fresh cocoons, which are taken out of a lot for testing purposes depends on the actual weight of the lot on offer. The cocoons on offer are divided into three batches.

— Batch weighing up to 1,000 kgs

— Batch weighing up to 2,000 kgs

— Batch weighing up to 4,000 kgs

The sample size of fresh cocoons taken out of each batch for testing is as follows:

— 1st batch - 2.0 kgs

— 2nd batch – 4.5 kgs

— 3rd batch – 6.0 kgs

In the case of dry cocoons, the quantities taken out for testing from each batch are as follows:

— 1st batch up to 400 kgs of dry weight – 0.8 kgs taken out for testing.

— 2nd batch up to 800 kgs of dry weight – 1.8 kgs taken out for testing.

— 3rd batch up to 1 600 and over - 2.4 kgs taken out for testing.

Drying of the Cocoon Test Sample

The cocoons received must be dried as soon as possible

and to an acceptable degree. To do this, the moisture content of the fresh pupa and the cocoon shell percentage must be measured. The percentage of drying is calculated according to the formula. Drying should be accomplished in one continuous process where the temperature is gradually decreased from 98°C to 60°C until the required ratio is obtained.

Evaluation of Eliminated Cocoons

This evaluation usually happens on a table under natural light. If natural light is insufficient based on time of day or weather conditions, elimination takes place under artificial light of 500 lux. to calculate the percentage of eliminated cocoons according to terms and conditions defined by cocoon classification, the following types of cocoons must be removed: double cocoons, thin-end cocoons, scaffold marked cocoons, malformed cocoons, flimsy cocoons.

Cocoon Batching

After elimination of the bad cocoons, the remaining cocoons are batched for the reeling test as follows:

Batch	*Sample cocoon (fresh weight)*	*Cocoon for reeling (dried weight)*	*Preliminary cooking*
	kg	grams	grams
1st	2.0	300 x 2	80
2nd	4.5	300 x 2	80 x 2
3rd	6.0	300 x 3	80 x 2

Cooking

A small sample of cocoons must be cooked to determine the correct cooking conditions for the specific batch. Once

Table 2. Cocoon classification systems of major sericultural countries

India	*Japan*	*Korea*	
Visual and mechanical test A. Visual inspection- cocoon shell weight (g/20 p.c.s. of bivoltine or 40 p.c.s. of polyvoltine) - 1 ~11 grade: 9.0-7.0 g (Jiangsu) 9.4-6.4 g (Sichuen) B. Mechanical test (Multi-ends reeling) - Length of non-broken cocoon filament (major item): 1~20 grades (950-340 m)- Percentage of good cocoons (auxiliary item): 1~ 8 grades(94-82%) - Size of cocoon filament and percent of insidestained cocoons (correction items)	Visual test $A = \frac{\text{Estimated Renditta constant}}{\text{shell \% of cocoon lot}}$ constants : 165-133 $\text{shell \%} = \frac{\text{weight of 100 shells}}{\text{weight of 100 cocoon}} \times 100$ B. Cocoon pricing cost of cocoon per kg. $= \frac{\text{Kakame cost}}{\text{Renditta}}$ - Kakame cost = silk price + income from by-products - (cost of manufacture + profits)	Mechanical test (Auto reeling) A. Cocoon testing items- Raw silk percentage of cocoon (%) - Percentage if eliminated cocoon (%) - Reelability percentage (%) B. Classification of cocoons - Grade : Reelability percentage5A : 100-85 4A : 84-80 3A : 79-75 2A : 74-70 A : 69-65 B : 64-60 C : 59-55 D : 54-50 E : 49-	Mechanical test (Auto and multi-ends reeling) A. Cocoon testing items - Raw silk percentage ofcocoon (%) - Percentage of eliminatedcocoon (%) - Length of cocoon filament(m) B. Grading of cocoon - Length of cocoon filament 10 class (33.5-42.5 point) - Cocoon reelability percent 10 class (43.5-52.5 point) Final grade (1+2) A : over 90 B : 88-89 C : 86-87 D : 84-85 E : below 83

these parameters are established, actual cooking of the entire sample batch can be completed.

Reeling of Cocoon Sample

Reeling should be carried out under the following conditions:

Item	*Multi-end reeling*	*Automatic reeling*
Temperature of groping end part	85°C	80°C
Reeling velocity	90 m/min	160 m/min
No. of reeling silk ends (per basin)	10 pcs.	3 pcs.
Length of croissure	10 cm	8 cm
No. of cocoons or objective size per reeling thread	8 pcs.	21 denier

Re-reeling

This should be carried out on large reels with a circumference of 1.5 m and the standard re-reeling speed of 160 r.p.m.

Calculation of results

The results of at least two or three reeling tests must be taken in order to calculate the classification. The resultant raw silk weight divided by the sample cocoon weight will indicate the raw silk percentage. The cocoon classification items are worked out by the following method:

Percentage of cocoon shell

$$\text{Percentage of cocoon shell} = \frac{\text{weight of cocoon shell (g)}}{\text{weight of whole cocoon}} \times 100$$

Points to be observed for testing percentage of cocoon shell are as follows:

- The percentage of the cocoon shell can be calculated from the weight of the whole cocoon and cocoon shell obtained from 200 cocoons from which 100 cocoons are taken separately.
- If possible, equal amounts of female and male cocoons should be selected for the sample.
- No defective cocoons should be included in the sample.
- As atmospheric conditions make the moisture content of the cocoons too variable, the sample should be selected from a point 10 cm beneath the surface of the batch of cocoons.

The average value of the percentage of the cocoon shell obtained from repeating the test twice must be graded within a 0.3 percent deviation. But if the difference between both samples is + 0.3 percent another test should be taken.

Estimated cocoon percentage

$$= \frac{\text{weight of e lim inated cocoon}}{\text{weight of sample cocoon (g)}} \times 100$$

The result should be expressed to one decimal place where the weight of sample cocoons is the sum of eliminated cocoons plus the weight of the good cocoons.

Length of cocoon filament

$$\frac{\text{Raw silk length (m) x Ave. reeling cocoon number per thread}}{\text{Total reeled cocoon number}}$$

Where, total reeled cocoon number
= Sample cocoon number – Converted carry over cocoon number

Ave. reeling cocoon number per thread =

$$\frac{\text{Sum of reeling cocoon number per thread}}{\text{Sum of reeling ends number checked}}$$

Length of raw silk is checked by the gauge. Sum of reeling cocoon number per thread is the total number of reeling cocoon verified 20 times during the reeling (this means once per unit work process). The length is based on the average of every reeling block. It is expressed by total number to the one decimal place. Converted unreelable cocoon number and carry over cocoons are calculated by converting them into length and expressed by full cocoon number to the one decimal place.

The cocoon number converted to full cocoon length

$$= 1.00P + 0.77H + 0.39M + 0.12L \qquad (1)$$

Where P : number of newly cooked cocoons but unreelable

H : number of heavy shell cocoons but unreelable or carried over

M : number of middle shell cocoons unreelable or carried over

L : number of light shell cocoons unreelable or carried over

The reeling cocoon number per thread is estimated down to two decimal places. The reeling work is not completed to the last single cocoon, but up to about 50 cocoons. The remaining cocoons are "carry over cocoons". They are divided into three kings: (H)...cocoons where only the outside layer has been reeled, (M)...cocoons reeled up to middle layer, and (L)...cocoons reeled up to inner side layer. These cocoons can be reeled along with other cocoons. That is why they should be converted into full cocoon length or weight and then the converted number has to be deducted from the total sample of cocoons to obtain the exact raw silk percentage or actual sample.

The converted number of full cocoon number is called "carry over cocoon number" which is calculated by multiplying the number of cocoons with Heavy layer, Middle layer and Light layer by cocoon convert indices. Also, the unreelable cocoons during the reeling or after reeling, are converted into full cocoon number by applying the same indices as for carry over cocoon. For example:

— Number of sample cocoons 340

— Length of raw silk 55 150 m

— Total cocoon number of snap check on reeling cocoon number per thread 720

— Total ends checked for reeling cocoon number per thread 91 ends

— Unreelable cocoons H(1), M(3)

— Carry over cocoons H95), M(14), L(21)

— Reeling cocoon number = 340-1-12 = 326

Where (1 – Convert unreelable cocoon number)

(12 – Convert carry over cocoon number)

Ave. reeling cocoon number per thread = 720 + 91 = 7.91

$$\text{Length of cocoon filament} = \frac{55150\ (\text{m}) \times 7.91}{326}$$

$$= 1338\ (\text{m})\ldots\text{batch(a)}$$

This is the length of cocoon filament in (b) batch is 1 349.1, the average of (a) and (b) batch is 1 343.6 m. Thus, the length of cocoon filament = 1 344 m.

Reelability Percentage

$$= \frac{\text{Reeled cocoon number}}{\text{Number of ends feeding}} \times 100$$

Where reeled cocoon number = number of sample cocoon – number of unreelable new cocoons + number of converted carry over cocoons

Number of ends feeding = number of cocoons fed + number of carry over cocoons – number of converted carry over cocoons

Raw silk percentage of cocoon =

$$\frac{\text{condition silk wight (g) + carry over cocoon silk yield}}{\text{weight of cocoons}} \times 100$$

Now, converted carry over cocoon silk (g) = weight of bave (g) x number of carried over cocoons to be converted to full ones.

$$\text{Here, weight of bave} = \frac{\text{conditioned silk wight (g)}}{\text{number of reeled cocoons}}$$

Where, number of reeled cocoons

= number of sample cocoons

– number of converted unreelable cocoons

+ number of converted carry over cocoons to full one

Raw silk percentage is estimated to two decimal places. The weight of bave is estimated down to three decimal places. The converted unreelable cocoons and carry over cocoons are based on the weight system, and the number of converted cocoons is estimated on the total to one decimal place. Cocoon classification is divided into 5 grades: A, B, C, D and E.

Cocoon exchange

The price of cocoons required for producing one kg of raw silk, (Cocoon Price Parameter [COO]), is calculated on the basis of 10 percent raw silk percentage cocoon to make 2A grade raw silk. The cocoon price parameter as a basis for estimating the cocoon price is obtained from the following formula:

(raw silk price + silk by-product income)

CPP / 100 kg cocoon =

$$\frac{\text{(cost of raw silk production + profits}}{\text{1 bale weight of raw silk (60 kg)}}$$

As the CPP by this method stands as the standard of C grade cocoons, if varied from it each 2.0% CPP per single grade is added to or deducted from the standard one.

Grade	A	B	C	D	E
% of CPP amended	+4.0	+2.0	0	-2.0	-4.0

Cocoon is computed from the following formula:

Cocoon price/kg = CPP ± amended CPP) x raw silk percentage of cocoon x (1-Discounting rate)

The resulting raw silk percentage is the raw silk test value multiplied by 0.9785, which equals fresh cocoon yield. The silk producer may have large stocks of cocoons in storage for reeling, which will result in the reduction of raw silk percentage. As the test sample does not match actual raw silk sold, consumed for testing in mills or in a "discounting rate" is used for raw silk during the reeling process of 2.15 percent. Where no cocoon classification system exists, a derived arbitrary raw silk percentage is used based on visual inspection.

A fictitious raw silk percentage is calculated using cocoon shell percentage, relative raw silk percentage from the previous year's assumed value. For example, in Table 3 when the CPP is 29 and the fictitious raw silk percentage of cocoon is 18 percent, the commercial price per kg is 29 x 0.18 = US$5.22.

Table 3. An example of cocoon pricing by visual inspection

Grades	*Shell % of fresh cocoon*	*% of good cocoon (%)*	*Fictitious raw silk % of fresh cocoon*	*Price of fresh cocoon ($ per kg)*
1st	22.5 over	98	18	5.22
2nd	21.5	97	17	4.93
3rd	20.5	96	16	4.64
4th	19.5	95	15	4.35
5th	18.5	94	14	3.19
Fair cocoon		Outside slight stained	11	3.19
Undergrade cocoon			8	2.32

6

Rearing the Muga Silkworms

Muga silkworm (Antheraea assama; family Saturniidae) is semi-domesticated and multivoltine. It is possible to rear 4-6 crops in a year. The rearing is out-door and exposes to environment conditions and attack of parasites, predators and diseases. It results in heavy losses due to weather conditions, pests, diseases and natural calamities. These losses are due mainly to the faulty selection of rearing places (Muga chung) and food plants, improper handling of worms, poor supervision of rearing, intensity of natural enemies, etc. Smoke from various engines and workshops, dust from gravel roads, some of the pesticides and herbicides used in agricultural fields and tea estates, D.D.T. spraying for eradication of Malaria are also some of the factors hampering the healthy growth of Muga-culture.

The essentials items for Muga rearing are as follows:

— Bamboo baskets,

— Mounting brushes (Kharikas),

— Secateur,

— Butterfly catching nets,

— Bow with pallets,

— Bird scaring instruments,
— Nylon mosquito nets,
— Wire-netted cages or Bamboo cages (chakaripera),
— Triangular bamboo shieves (chalani)
— Alluminium pans
— Bamboo poles,
— Gamaxin,
— Tugon bait,
— Gum,
— Lali, molasses,
— Dry leaf Jalis, etc.

All rearing appliances must be disinfected with 4 percent formal· dehyde solution in time. The life cycle of Muga may be divided into four stages, viz.

— Egg,
— Larva,
— Cocoon (pupa) and
— Moth.

A suitable rearing place (Muga chung) is to be selected. It should not be low lying and shady. Medium size full-grown Som/Suala trees with more branches and fresh leaves are to be selected for rearing. Early stage worms should preferably be reared on dwarf plants for reducing loss of worm. Tall trees will reduce the production. It is also difficult to hang "Kharikas" and to transfer the worms to other trees after exhaustion of leaves on tall trees.

Worms also find it difficult to crawl up the tall trees in search of suitable leaves. It increases the mortality of worms of course, flies wasps etc., may not be able to attack the worms easily in tall trees as they can fly to a particular level only. The rearing place (Muga chung) should be cleaned including the base of all trees to avoid pests. The dry and diseased branches, twigs and leaves, nests of ants and wasps, etc, are also to be removed before starting the rearing.

Plants infected with ants, aphids, wasps, termites are to be avoided for rearing. A band of straw should be wrapped around the trunk of each tree at a height of about one meter. This will act as a barrier for the worms from coming down and prevents ants, etc., from going up. Banana or pine apple leaves may be put above this band so that worms cannot come down after exhaustion of leaves. These will facilitate to collect them easily. Ashes, lime and sand mixed with kerosene may be put at the base of trees.

Gamaxin may also be sprinkled at the base of trees to prevent ants but it must be done at least before a fortnight of mounting the worms. Fish refuse, mollases etc., may be kept at the base of trees to attract the ants. Ants should be burnt regularly in the morning and evening. The nests of ants/wasps from nearby bushes/grasses are also to be cleared. It is better if the apical buds in the branches may be removed before two weeks of rearing as buds and every tender leaves are not suitable for rearing. Insecticides/fungicides should not be sprayed before rearing.

The eggs of muga silkworms are brownish in colour. They hatch out within 1 to 12 days of eviposition during summer and a16 to 25 days in winter. The worms start hatching from early in the morning till about 10:00 a.m. under normal condition. Hatching will be ever within 24 days. Newly hatched worms eat their egg shells. Worms hatched out during first two days are to be considered for farm rearing as they graw uniformly and are healthy. Larvae hatched up to 4 days are to be kept for commercial rearing.

The percentage of hatching is 40% to 95 according to the season. Hatching will delayed during summer by keeping the eggs at 6°C to 10°C and hastened during winter by exposing them to 28°C to 30°C as experimented at Muga Research Station, Dhakuakhana (Assam). Eggs kept in controlled condition and incubated eggs will hatch uniformly and worms would complete maturing with 3-4 days. Eggs should be kept in perforated boxes in an disinfected room and ants, lizard, etc. should not be there. Eggs with kharikas (bamboo orthatch) or the loose eggs in bamboo/paper baskets are to be tied with or hang on suitable leafy branches and on the trunk when the majority of the worms are hatched out (4 A.M. to 10 A.M.).

The newly hatched tiny worms will crawl up by themselves to the leave of Som/Suala by their natural instinct. They start eating from the margin of suitable leaves. They eat up the midrib and even the petiole of the leaf during the last two instars. Another method of mounting the newly hatched Muga worms is that small twigs bearing soft and tender leaves are lightly put ever the newly hatched worms on the "Kharikas" or in the egg boxes/baskets. They will crawl up to the leaves.

The twigs with the worms are then tied with suitable branches at different places of the Som/Suala trees for uniform distribution. They are to be put either on the east or north side of the trees to get sufficient sunshine by the help of bamboo poles. The suitable time for mounting the newly hatched worms is morning. It is not advisable to mount during heavy showers, heavy wind or storm. It is better if a limited number of worms can be mounted on a particular tree, where leaves will be sufficient upto maturing to avoid frequent transfer of worms. 5 layings may be a reared in a full grown Som/Sualu tree and about 200 cocoons may be harvested in one crop.

Overcrowding by mounting more worms on a tree by the rearers compel them to eat hard leaves unsuited to their age. It is the main cause of flacherie desease of Muga. It is desirable to mount the worms hatching in different days on separate plants. Kharikas/bamboo baskets are to be collected, disinfected with 2% formaldehyde solution and preserved properly for future use. The younger worms prefer soft and tender leaves and matured leaves at the advance instars stages. It has been observed that the rate of feeding is higher at night than in the day time. They take rest (about 5 minutes) after every stretch of feeding.

Healthy worms take the whole leaf even soft twigs from the top of the tree. Larvae move much during the first instar-1 stage in search of the suitable leaves than during the other instars. If and when change of food plant is necessary, worms should be mounted first on Sualu and then transferred to Som. Worms find difficult to eat Sualu leaves due to presence of gall wasps and of flowers. Likewise, fly pests are more in Digloti rearing and the fruits of Digloti also hamper the worms to a great extent.

Muga worms are green in colour with thorny warts. They shed their skin (moult) four times their larval period. Worms move more before moulting to find out proper place for shelter. Moulting period varies from 24 to 36 hours depending upon this season of rearing. The larvae eat their cast off skin. No disturbance and transfer of worms should be there during moulting. The transfer of worms after exhaustion of leaves should be done after 3-4 hours of moulting (ecdvsis).

Day and night constant and close watching is indispensable to avoid pests and to scare birds, bats, etc. This will also help to assess the condition of the worms. Dead worms should be collected twice daily in the morning and evening. They are to be burnt and buried away from the rearing place. Microscopic examination of dead and diseased worms should be done to detect diseases. It is advisable to conduct rearing of unhealthy and weak worms separately on fresh plants.

Muga worm start descending of their food plants when leaves are exhausted. They are picked up, sorted out according to their stages and hung on few leafy branches of other suitable tress with the help of triangular bamboo shieves (chalani) and bamboo poles. Another method for transferring the worms is keeping leafy twigs alongwith the worms may be put on other suitable trees with the help of bamboo poles. One should handle them carefully to avoid injury to worms by forcible detachment.

Worm falling from the trees should be picked up by some suitable leaves (branches). Worms defecate orange shaped litters with six longitudinal furrows. Worms usually hide on the under surface of the leaves and have a strong gripping power. They are very sensitive and disturb easily. They stop eating leaf and shrink their body (expressing their anger) when disturbed by movement, sound or touch etc. If removed by force, they vomit a liquid. Rough surface (bark) of the trunk of the tree should be made smoth. It will help in picking up worms easily without damage during transfer. Smaller worms would be picked up first and the bigger worms later. The blue colour larvae are bigger and heavier than the normal green colour larvae.

The larval period varies from 22 to 50 days according to the season of rearing. At the end of fifth instar the worm attains maturity, stops feeding and empties its alimentary canal by passing out the last excreata (green semi-solid) mass followed by coloureds slimy substace). Worms consume about 20 per cent leaves during first instars and 80 percent during the last instar alone.

Matured Muga worms become slightly smaller, flabby and transluscent and make a hallow sound on touch. They come down from the tree in the evening due to effect of negative geotropism. It makes the collection of worms easier for the nearers. They are collected by the rearers in bamboo baskets quickly and may be put on 'Jalis' directly. The effective rate of rearing varies from 20-60 Cocoons per laying, depending on the season rearing.

Mugaworms feed on Mejankari and chapa leaves produce creamy white silk. It fetches a higher price. Of course, the growth of Mugaworm on Mejankari is said to be slow and unhealthy in comparison to that of som Plants. However, rearing on Majankar may be conducted where it is growing abundantly for producing creamy-white 'Mejankari' silk.

Generally a crop completes maturing within 3-4 days. Of course, some crops take upto two weeks to complete maturing. The worms maturing earlier produce male moths, those that mature later produce more female moths and worms maturing in between (Bhor-pok days) produce an approximately equal number of male and female moths. The matured worms are then taken to the seed preparation house or any other suitable place and put them into the Jali prepared earlier from suitable dry leafy twigs and kept hanging vertically in dry place for spinning cocoons after being counted.

The number of worms in a 'Jali' are to be adjusted according to the space available in it. 'Jalis' may be arranged on bamboo mats where mutured. The mature worms are then taken to the seed preparation house or any other suitable place and put them into the Jali prepared earlier from suitable dry leafy twigs and kept hanging variety in dry place for spinning cocoons after being counted.

The number of worms in a 'Jali' are to be adjusted according to the space available in it. 'Jalis' may be arranged on bamboo mats where matured worms are kept and they will crawl up to the 'Jalis' easily. Some one should observe the worms atleast for an hour after hanging the 'Jalis' as some worms way fall down from the 'Jalis' while searching for suitable place for cocooning. Those worms are to be kept on 'Jalis' again. Sex in larval stage can easily be distinguished by the external genital markings.

Four dots in case of female and a V-shaped marking the in male worms are seen in between the 8th and 9th segments on central side. Sex markings in pupa are more prominent than in worms. Separate cocoonage should be used for males and females in case of production. 'Jalis' are to be prepared from the leave of suitable trees. The quality, size, colour, etc of the cocoons differ according to the food plants on which they feed. It is said that the age of the tree has a great effect on the colour of the silk produced. The cocoon spun by Muga worms feeding in Diglati leaves is smaller to that of other Muga food Plants.

Cocooning

Cocooning is completed within 3-4 days in summer and 7 days in winter. Pupation is completed after 15-30 days of cocooning. No disturbance should be there to the Jalis up to a week during cocoons formation. The female worms spin bigger cocoons than the males. Usually cocoon formation starts during the day time. Cocoons are harvested normally after a week of spinning during summer. Of course, it is delayed up to 1½ weeks during winter. The normal and the inferior cocoons should be harvested separately. Cocoons are cleaned of the adhering leaves at the time of harvesting. Good and healthy cocoons are kept for seed production, and the others for reeling and spinning purposes. The weight of cocoons, silk shell, pupa and the length of filament varies according to season.

Indoor Rearing

The indoor rearing for the whole larval period may be practiced for producing basic seed cocoons and for maintaining the parent stocks. It may be done by bringing leafy branches or Som/Saulu trees and putting them in some container (non corrosive) having cold water. These branches are to be changed from time to time whenever necessary to keep the freshness of the leaves.

It should be done after proper experimentation for better harvesting. It may result about 95 per cent success of the cocoon crop. It does not require regular watching like that of outdoor rearing, reduces expenditure, loss of worms and mortality. It also protects the Muga worms from their enemies like ants, wasps, birds, bats, etc. Continuous indoor rearing for many generations throughout the larval period may weaken the strain.

So, indoor rearing up to second moult or third stage and outdoor rearing for the remaining instars preferably under net covering with proper care will help to maintain the health and vigour of the race intact and in harvesting more cocoons. Muga worms like to take water drops accumulated on the leaf for which sprinkling of water once a day on the twigs is advisable in case of indoor rearing. Windows should be kept open from time to time as required. Sericultural hygiene must be followed.

Disinfection is to be done to the rearing houses used for indoor rearing of Mug worms and also the rearing appliances like that of Eri and Mulberry silkworm rearing houses. Chowki rearing especially during summer may be adopted for better rearing results. It is not possible to conduct large scale rearing of Muga worms indoor. The outdoor rearing for the whole larval period also increases loses to the commercial rearers. So, after rearing up to third instar indoor, worms should be transferred to dwarf leafy trees for outdoor rearing, for the remaining larval period.

Dwarf plants covered with fine nylon mosquito nets or Wire-netted cages for outdoor rearing of young worms at least up to the second moult will also save the worms to a great extent from ants, wasps, flies, etc. Worms hatched out from the eggs of different mother moths are to be reared on separate trees without mixing them Cocoons harvested from such rearing should be kept separately. Male and female moths emerging from these cocoons are to be allowed to pair disallowing brother sister mating (inbreeding).

It facilitates in keeping the hybrid vigour in subsequent generations. It will help in the maintenance of the purity of the Strain without deterioration. Further research is necessary in evolving and developing some suitable interspecific hybrid for the greater interest of the Muga silk industry. Yellow body mutant of Maga should also be tried for outbreeding which would help in maintaining vigour and health of the worms. The body weight of Muga worms is abouts 0.008 gms to 5.5 gms from 1st instar to 5th instar. The weight of cocoon varies from 4.5-6.0. gms and the shall weight from 0.30 gms to 0.60 gms.

Rearing Season

The rearing of two major commercial crops. (in four batches) a years may suitably be practiced during spring (early and late) for producing reeling cocoons. The may be adjusted according to the local conditions. It is better to conduct rearing of summer crop in cold climate (at high altitude) as mortality and flies, wasps, etc, are more during summer in hot places. It will help in supplying the seed cocoons for the autumn crop. It will also provide hill amelioration to the worms.

Maximum possible layings should be reared during springs (first crop and autumn third crop) rearing as these are the two most suitable seasons for rearing of Muga. It will increase the quantity and the quality of cocoons. Necessary steps must be taken to control different diseases and to get rid of various ants, wasps, flies, birds, bats, etc., for harvesting a successful cocoon crop. Tugon bait, gum, lali, molasses, etc. may be used in alluminium pans for controlling and killing ants, wasps, flies, etc., Birds scared away by bows and pallets and other bird scaring instruments.

Most of the wild Muga worms (yellow, blue and orange in colour) are bivoltine and do not emerge during winter. The collection of wild Muga seed cocoons from different places of hills and plains of Meghalaya and Assam and also from certain places will help the rearers in improving the production of quality cocoons by hybridising them with semi-domesticated strains. It is essential to multiply the hibernating strains of Muga worms as such breed will eliminate the hazardous winter rearing. They are vigorous and stronger and will solve the problem of seed cocoons to a great extent.

Rearing of selected strains may be conducted in suitable green/glass houses (specially early spring crop when hail storm, pests are prevailing), under controlled temperature (24°C-30°C) humidity (75%-85%) and light. Suitable selected clonal plants on tubs may be used for the purpose for making nutritious leaves available. This may also help in reducing degeneration of Muga worms.

Rearers usually do not select the trees in advance before taking up rearing for which leaf quality varies and affects the growth of Muga worms. Most of the activities of Muga culture happen to be in darkness such as the maturing of worms, emergencae of moths, pairing of moth, and laying of eggs. So, the Muga rearers will have to work even at night.

About 300 Som/Sualu Plants may be planted in one acre (about 3 bighas) of land with a spacing of 3m x 3m. In this plantations about 900 layings of muga worm can be reared in one crop easily. Such, major crop for producing reeling cocoons may be taken up twice at year during April-May (till mid June) and September-October (up to middle of November). Thus a family having three adult members may rear 1800 layings) in two crops and harvest about 72,000 cocoons(40 cocoons per laying) within a period of three months (two crops).

The income from the rearing thus would be Rs. 14,400.00 by selling 72,000 reeling cocoons at the Rs. 200/- per 1000 cocoons. Muga rearing is done generally by the rearers themselves and no paid labourer is engaged. So whatever amount is obtained by disposing the cocoons may be considered as income of the family. The capital investment is also reasonable, except counting the wages of their self-employment, cost of seed cocoons/seeds and rearing appliances, the plants as they can rear more layings.

Inter-crops will also increase the income of rearers to some extent. Indoor rearing for early stages worms followed by outdoor rearing in later stages will also help in harvesting more cocoons (100 cocoons per laying). It is advisable to conduct the rearing of two major crops in two batches each (early spring, late spring and early autumn, late autumn) for accommodating more layings and to ensure harvesting of more cocoons.

Muga silk industry may be developed in rural areas where their food plants are available in nature. They may also be planted in vacant places and in waste land. The farmers having small land-holding may take up it easily. The people of Muga food plant growing areas are to be educated to preserve them for their own benefit. They should also be taught the improved rearing technique of Muga worms.

Seed Production

Seed production is the most important aspect of Muga silk Industry and the success of rearing mainly depends on it. Seed production may be divided into four divisions, viz. (A) Selection of seed cocoons, (B) Preparation of seeds, (C) Seed examination and (D) Seed supply.

The seed preparation houses should be spacious. They are to be provided with a microscope, heater, lights, air cooler, humidifier, incubator, refrigerator, sprayer including formaline, etc. They should be free from rats, lizards, spiders, ants, etc. Important characters for selection of Muga are viability, fecundity, larval period, larva weight, mortality, disease resistance, cocoon weight, shell weight, filament length, demier, etc. Selection may be done considering all required characters of the Muga silkworm (chromosome number n=15) through out its life cycle.

Seed Cocoons Selection

Better seed cocoons are always required to produce better silkworm seeds. If possible, it is better to select them during the larval stage. There are hibernating and non-hibernating Muga strains. We should try to increase the production of hibernating Muga broods. Experienced Muga rearers visit the Muga rearing in the last stage before maturity. They observe the worms, the mode of their eating leaves on the tree, visual examination of some of the worms, percentage of mortality, etc.

Healthy worms usually eat the leaves from the top of a tree and even the twigs. They are bright green in colour. Their dry litres powder when crushed, 2-3 littres are present at the rectal portion of a healthy worm. They show their sensitiveness without vomiting upto 4-5 times, while touched. Seed cocoons are usually selected and graded on visual examination through long practical experience, after the harvest. Seed cocoons should be selected from and around "Bhorpok" (mid-maturity) stage to obtain approximately equal number of male and female moths for seed preparation.

Healthy, well built, robost cocoons with live pupae, fine denier silk, more percentage of silk content, good reelability, longer filament length, compact in texture, without any pest ialso have life like that of human beings. It is better to carry seed cocoons along with the 'Jali' to avoid damage during transportation. It is advisable to collect healthy disease free seed cocoons from a distant place so that change of food, climate, etc, occurs and better cocoon crop may be obtained. Selected seed cocoons are to be kept in rat-proof well ventilated rooms in wire-netted and bamboo cages (chakaripera) in one layer.

More space is required for preservation of Muga seed cocoons. Seed cocoons may be kept in garland when a large number of Commercial seeds are to prepared. Each garland should have 50-150 cocoons Seed cocoons received from different areas are to be kept separately Seed cocoon should not be stored in damp, poorly ventilated and in dark places. Wire-netted cages are to be used to maintain the purity of the strains of Muga, especially where parent stocks are maintained. The cages may bae arranged in tiers for economy of space.

The four legs of the lower most cage should be kept on ant-wells. Preservation of spring and autumn breeds seed cocoons should be in cold storage for delay and regular emergence of moths. The period of diapause is in pupal stage. This will also help in controlling pebrine disease. Such preservation of seed cocoons will help in avoiding summer and winter crops. This will ensure the seed supply in autumn and spring rearing. It is better if the seed cocoons may be preserved under regulated temperature and humidity followed by proper incubation for uniform emergence.

Seed Cocoons may be preserved in cold storage at 25°C first and then to reduce the temperature to 5°C after every 12 hours upto 5°C Seed Cocoons may be preserved upto 4 months. Similarly while releasing these preserved seed cocoons the temperature should gradually be increased by S°C after every 12 hours and then to keep at the room temperature.

Seed cocoons also may be preserved in high altitude during summer with proper care in transportation.

Necessary steps are to be taken to control the life-cycle of Muga silk-worm by artificial means subjecting the pupae to hibernate for avoiding the summer and the winter rearing. It may be induced by controlling the photoperiod at the larval stage. The photoperiod effect combined with temperature and humidity renders some Muga-worms to go for diapause and others to continue with the voltinism

The price of seed cocoons is too high but the rearers purchasing capacity is limited. So, marketing of seed cocoons should be regulated. Rearers co-operatives may also be formed to avoid exploitation by middlemen. Introduction of a crash programme for production of more muga seed cocoons would also ease the position.

Seeds Preparation

Muga seed preparation is time consuming and hazardous task. Usually for cocoons are required for preparation of a single laying. Moths emerge from the seed cocoons after 2-4 weeks from the date of cocooning according to season. In normal condition the emergence of moth takes place from evening till midnight. Male moths emerge earlier than the female moths. They can easily be identified. The male moths are copper brown to deep brown in colour while the females are yellowish light brown. The males have broader antenna and narrow and small abdomen. The females have smaller antenna and large abdomen.

Seeds cocoons are to be refrigerated at 5-10°C in early pupal stage and incubated at 34°C to synchronise the time of emergence of moths of different strains of Muga for hybridisation. Moths may be refrigerated for 3-4 days without adverse effect. Male moths may also be refrigerated for about a week. They may be utilised for a second time pairing, if and when necessary. Refrigeration for a longer duration is not advisable. Moths are to be selected after emergence considering their health; natural brown colour with thick wings, without any deformity, and urine like that of pure milk, etc. The male and female moths are allowed to pair in the cage naturally. Some of them may not pair for which mechanical means, such as, moving them by hand and blowing air by mouth for early pairing are to be adopted by the breeder.

Maximum care must be taken to avoid damage to the female moths. Unpaired female moths may also be kept outside in a safe place tying them on "Kharika" for pairing with wild males after darkness in the evening and should be collected in the early morning of the following day. Moths like darkness for pairing with comparatively lower temperature and high humidity. Male moths are more active and smart fliers. Female moths are passive and generally do not fly. The female moths are tied by one of the hind wings and fastened to a "Khasika" first and then allowed them to pair. Paired moths are taken out from the cage carefully and the female ones are tied, unpaired and kept for egg laying by hanging the "karika" and strings/wire arranged already for the purpose.

Moths are usually allowed to pair for about 12 hours due to practical reason though 4-5 hours are sufficient for fertilisation. They start pairing in the evening and are unpaired in the following morning by hand. One must take care at the time of unpairing so that female moths will not be injured. Application of light or heat will help in separating couples by themselves. A single male moth can be used for a second coupling without affecting fertility and egg laying when male moths are less.

After unpairing, the female moths are kept for egg laying on "Kharikas". They prefer darkness for egg laying like that of other satuniids. They start laying eggs within

12 hours. They are allowed to lay eggs for 3 days, which are considered for rearing purpose. Eggs laid latter are less viable and the worms become weak, and development is poor. A female moth usually lays 150-300 eggs. Muga moths lay maximum eggs in spring and autumn, and minimum during summer.

Eggs adhere to "Kharika" or in a small bamboo/paper basket with cover for basic seed preparation to help in proper examination. More moths may be allowed to lay egg together in longer bamboo Kharikas for large scale seed production. The moths with Kharika should not be exposed to sunlight and rain. Likewise, more unpaired female moths may be kept in paper/bamboo basket for preparation of large scale loose eggs but it is not advisable as it would be difficult to examine the mother moths.

Sometimes smoke or light is to be introduced in the room as shock treatment to the female moths. It gives better effect in eggs laying. Deposition and preservation of eggs should be conducted in a cool room with sufficient humidity. Incubation may be practiced for regular and uniform hatching at 40°C. It will also help in controlling pebrine disease. The moth is a non-feeding stage and dies within 7-15 days after emergence. Continuous inbreeding will lead to degeneration of the merits of the race and also reduces vigour.

Hybridisation (out-breeding) in between two different group of Muga is essential for commercial rearing. Domesticated Muga may be hybridised with the hibernating wild Muga from time to time to give them hybrid vigour and to avoid the troublesome summer and winter broods. Muga silkworm is a primitive insect and has no means of large scale hybridisation in absence of related strains. Government farms should maintain parent stocks of different strains correcting from various places to study their behaviours and inter or intra-specific hybridisation for keeping hybrid vigour. It would help to prepare hybrid seeds in subsequent generations.

Survey of potential seed producing areas and collection of several strains from different area are necessary. These may be reared in small batches in a farm and all the cocoons harvested from such rearing may be mixed allowing them to have random mating which will give hybrid vigour. An alternative to this, these strains may be reared separately in different farms and only male cocoons are to be exchanged to prepare hybrid seeds among them.

Seed Examination

Visual examination play an important role in seed examination one may select better layings through his long practical experience. Eggs with glue bigger in size, laid in clusters with fibrous body dust and brown in colour are said to be healthy. Weak and dead moths irregular layings, less number of eggs in a laying, under developed and unfertilised eggs, etc, are to be rejected before seed examination. This is to be followed especially for parent stock maintenance. It will also help in quick microscopic examination. It is advisable to conduct examination in all the stages of the life-cycle of the Muga silkworm. Microscope, glass slides with cover, mortar and pestle, scissors, caustic potash solution, etc., are essential for seed examination.

After egg laying for three nights, the abdominal portion of mother moths are cut by scissor and crushed with the help of mortar and pestles, sprinkling a few drops of distilled water or ½ drop of 2 percent caustic potash solution. Then a drop of the suspension is placed on glass slide by stick/rod covered with glass cover and examined under microscope. The eye-piece, objective, etc. of the microscope should properly be cleaned and accurate focusing must be there for correct microscopic examination. It is better to examine a mother moth at a time and in case of basic seeds it is a must. In case of large scale industrial seed production, examination of ten moths taking at random from a particular lot is to be examined at a time.

The organisms that cause diseases of Muga silkworms are virus, bacteria, fungus and sporozoa. Generally, the eggs laid by moths having mild viral, bacteria or fungal infections may be use for rearing as there is difficulty to get sufficient seed cocoons. But we should not consider any when they are infested by sporozoa (pebrine), which is transmitted from mother to the off-springs, through eggs. Such infected mother moths and their eggs must be destroyed without delay by burning and burying them at a distant place.

Loose eggs are to be collected on the fourth day of oviposition after examination of mother moths and washed (disinfected) with 2 percent formalin. Washing of eggs for about 5 minutes with 2 percent formaldehyde solution followed by washing in clear cold water and drying in shade is essential to avoid contamination of diseases before supplying to rearers. Heat treatment for pebrinished pupae and eggs may be done like that of oak tasar after proper experimentation. Disinfection with Kharika is more advantageous.

Disinfected eggs are to be kept without overcrowding in cool and moist place. Hatching may be delayed for few days by refrigerating the eggs if and when necessary after 48 hours of oviposition. But is should not exceed more than a week for normal hatching. Loose eggs after disinfection may be preserved in a box having an arrangement for light on the top and for air on the sides. Eggs in boxes may be incubated at a required temperature and humidity which will help in uniform hatching. The eggs inside the box should be kept in one layer and not in heaps.

Seed Supply

The process of transportation of seeds has a direct effect in the cocoons crops. Every one must remember that there is life inside the eggs. No abnormal heat and direct sun should be experienced by the eggs during transit. Otherwise, it may result the failure in rearing, especially as it causes mortality in the 5th instar before maturity. One must realise that the embryo is being developed inside the egg like that of the development of the baby in the mother's womb. Seeds should reach the destination within the seventh day of oviposition. A well-perforated wooden box with a thin cloth lining inside is to be used for sending seeds (loose-eggs) with proper operation. It is advisable to transport eggs on 4th/5th day of incubation.

Quantity of eggs is to be put inside the box according to its capacity and in no case it should be over loaded to avoid damage to developing embryos. Eggs may be supplied along with "Kharikas" for nearby places. Eggs are to be transported preferably in cooler hours. Tender leaves may be put in the eggs carrying boxes, if hatched eggs are transported. It is better to supply seed cocoons with "Jali" instead of seeds wherever and whenever possible with proper care in transportation. Seed cocoons should be transported carefully after 6-9 days of spinning.

Government should take appropriate steps to supply muga seed cocoons and disease free seeds to the rearers at reasonable cost in required seasons. Rearers usually cannot continue the rearing for more than three generations of a particular lot of muga due to successive inbreeding. Field Staff should arrange for microscopic examination of mother moths in rearers' house to get disease free eggs.

7

Silkworm Genetic Resources

The domesticated silkworm species, *Bombyx mori* L. evolved almost 4600 years ago from the wild species, *Bombyx mandarina* Moore, which is a native of China and Palaearctic region. The eggs of silkworm, *B. mori* were first introduced from China into Japan and Korea in the first century and subsequently into Middle Eastern and European countries and later into the neighbouring countries around China in the sixth century. The historical background of silkworm entry into India is still a mystery; and the historical evidence indicates that a flourishing silk trade was practising between India and Rome/Greece during Kaniska period.

The rich tradition of silk and silk use are evident from ancient sacred literature like the Rigveda, the Ramayana, and the Mahabharatha, which are more than 2000 years old, but the information about indigenous silkworm races and their stock maintenance are not well documented. Silkworm rearing was prevalent in Kashmir and North Eastern states during sixteenth century, the Moghul period where the univoltine and multivoltine silkworms were respectively reared and the Tippu Sultan introduced silkworm rearing in south India in 1875.

During eighteenth century, the British rule in India, quite a few univoltine and bivoltine races were imported from Italy, France, Russia and China, and the races were bred and maintained by the farmers; and there was no systematic maintenance of the silkworm germplasm and hence only few races survived under Indian climatic condition. At present only few old indigenous races are surviving viz. Barapolu, Chotapolu, Nistari, Sarupat, and Moria, whereas the indigenous univoltine Kashmiri races are almost extinct.

Systematic silkworm stock maintenance and breeding started in the early nineteenth century. Prior to 1922, only pure races were reared and hybrid silkworms were introduced later, Pure Mysore × C. Nichi was probably the first hybrid in Karnataka and exploitation of hybrids in West Bengal and Kashmir came much later during 1956 and 1959 respectively.

Silkworm genetic stock maintenance started during 1940 in an organised way at Sericultural Research Station, Berhampore in West Bengal and subsequently temperate silkworm germplasm stocks were established at Univoltine Silkworm Seed Station, Pampore in Kashmir and multivoltine and bivoltine silkworm stocks were established at Central Sericultural Research Institute, Mysore in Karnataka and Coonoor in Tamil Nadu.

During the recent years, biodiversity conservation programmes have drawn the attention of many countries including developing nations, because of the genetic erosion due to indiscriminate use of bio resources and damage to the environment, destruction of forest, human interference in eco-system, upsetting the equilibrium of the biosphere. The Convention on Biological Diversity (CBD) organised by United Nations Conference on Environment and Development (UNCED) at Rio de

Jeneiro Earth Summit in 1992 made an awakening call to draw the global attention for conservation of biodiversity.

Since then the biodiversity conservation and gene bank maintenance have gained greater momentum since the germplasm resources are considered as "Common Heritage of Mankind" and "Sovereign Right of Nations". The issues related to access the genetic resources and its sustainable use, benefit sharing, farmers rights are being deliberated at various national and international fora.

Realising the importance of biodiversity conservation for sustainable development of agriculture, the Consultative Group on International Agricultural Research (CGIAR) established the International Board for Plant Genetic Resources (IBPGR) in 1974 at Rome with a global network of genetic resources centres, mainly for conservation of natural genetic resources including the wild species to promote crop improvement programmes and increase the food production. The role of wild relatives and wild species in agricultural crop improvement are well known. Similarly, there is an urgent need for seribiodiversity conservation, particularly the wild relatives of *Bombyx* and *Bombycidae*.

Improvement in silkworm race heavily depended on the geographical races of *B. mori* and the wild relatives of *Bombyx* were not explored, unlike in agriculture. Whereas in agricultural, horticultural and sericultural crop improvement programme the wild species of several crop plants have contributed very valuable genes for resistance to diseases and pests and tolerance to adverse agroclimatic conditions and similar exploitation of genes from wild relatives of *B.mori* have not been reported.

The genus *Bombyx* Hubner has two species, *Bombyx mori* L. and *Bombyx mandarina* Moore. Apart from the genus *Bombyx* there are eleven other genera in the family

Bombycidae Hubner; 1) Genus - *Theophila* Moore, 2) Genus - *Ocinara,* 3) Genus - *Mustilia,* 4) Genus - Gunda, 5) Genus *Penicillifera* (Walker) 6) Genus - *Ernolatia* (Moore) 7) Genus - *Norasuma* Moore 8) Genus - *Trilocha* Dieri, 9) Genus - *Prismosticta* (Swinhoe), 10) Genus - *Andraca* (Walker), and 11) Genus - *Ectrocta* (Hampson). Among these genera, *Theophila* and *Ocinara* are very close to the genus Bombyx.

The wild sericigenous species of *Bombyx, Theophila* and *Ocinara* are naturally distributed in the Himalayan ranges of Indo-China range and also in Andaman Islands in India, besides, Jawa, Sumatra, Borneo and Malaya Peninsular. The wild species of these genera have not been explored for transferring the useful genes to confer resistance to diseases and tolerance to adverse agro-climatic conditions into the domesticated species, *B.mori.* The useful genes from the wild relatives of *B. mori* may be cloned and these cloned genes may be transferred into the germ cells of the silkworm to develop transgeneic silkworm.

Hence, there is an urgent need to collect and conserve the wild species of *Bombyx, Theophila* and *Ocinera* and study their genetics for possible use in the breeding programme of *B.mori* and widen the genetic base as well. Indian gene centre is harbouring great faunal diversity and nearly 11.9% of the world flora are present in India and hence recognised as one among the twelve mega biodiversity rich centres of the world.

Floristically India is very rich, harbouring three mega centres of endemnism i.e. Western and Eastern Himalayas and Western Ghats. It is a treasure house of several diverse sericigenous flora and fauna. Wild species of *Bombyx* and other genera of *Bombycidae* do exist in the great Himalayan ranges and Andaman islands, under natural habitat and therefore the Indian gene centre possesses a rich seri-genetic resources.

Eggs and cocoons of a wild silkworm belonging to Bombycidae were collected from wild mulberry tree *Morus serrata* near Kedarnath (30.47°N, 79.02°E) at an altitude of 800 meter above MSL. The eggs were incubated and rearing was conducted on the mulberry plants at Central Sericultural Germplasm Resources Centre (CSGRC), Hosur and the produced cocoons and eggs are very similar to *B. mori*. It is a potential and interesting genetic material with several unique characters, utilising such wild relatives of *Bombyx,* it is quite possible to create additional seribiodiversity and widen the genetic base of *B. mori*.

Biodiversity is the result of evolution that is a continuous phenomenon induced by natural selection pressure and the population of organisms evolve through adaptation to the biotic and abiotic stress. Ever since *B.mori* was domesticated, the species does not survive in the wild state in natural condition and also does not survive without human care and hence natural selection induced genetic diversity in *B.mori* is rather very limited to voltinism.

Hence, it is very essential to conserve and utilise the wild relatives of *Bombyx mori* to broaden its genetic diversity, apart from the geographical races, mutants, sex-limited races, evolved breeds and breeders genetic stocks. The wild relatives of *Bombyx* are very vulnerable and the vulnerability at different spatial and temporal scales are not known. The design of biodiversity network in sericulture involving the complementarity of wild relatives and domesticated *B. mori* is also not well established. Therefore, conservation of wild as well as domesticated seribiodiversity resources is very essential for sustainable development of sericulture since loss of genetic resources of domesticated and wild relatives of *Bombyx* species along with their unique genes may disadvantage future generation.

Conservation of Silkworm Resources

Prior to the establishment of the exclusive centre for sericultural germplasm, the silkworm germplasm activities were carried out at the three Central Sericultural Research & Training Institutes (CSR&TI) at Berhampore, Mysore and Pampore, few Regional Sericultural Research Stations (RSRS) at Jammu, Dehradun, Kalimpong and Coonoor; some of the Universities and state funded sericultural research institutes. After the establishment of the germplasm centre at Hosur, the entire base collection of silkworm germplasm was collected periodically from these institutes and universities and deposited as base collection in the national repository for further conservation, maintenance and utilisation.

Development of Silkworm Germplasm Resources

Collection: Eversince silkworm rearing was introduced into India, several silkworm races were brought from China, Japan, Russia, France, and Italy at different times and few of them survive and naturalised in India. Also several new races were evolved in different regions and maintained at local conditions. Therefore, the CSGRC, Hosur, has a regular programme to collect the available natural variants, geographical races, and wild species of Bombyx through exploration, survey, contacts and exchange. The centre has so far collected 357 silkworm germplasm from different geographic regions along with passport data.

In India, the Central Sericultural Germplasm Resources Centre (CSGRC), at Hosur (12.45°N, 77.51°E; 942 m above MSL) maintains 357 silkworm germplasm accessions, out of which 63 are multivoltine and 294 are bivoltine silkworm accessions. The silkworm germplasm available in India are of three categories viz. i) indigenous, ii) introduced, and iii) evolved accessions. The Indian collection of silkworm germplasm has representation from 12 countries besides native Indian germplasm

The silkworm germplasm available in India has more of evolved races and there is a low representation of geographical races and races with sex limited markers, which indicate that there is great scope to enrich the silkworm germplasm from other geographical regions, particularly from Europe and Asia. The present day silkworm genetic stocks maintained in India include collections from different regions of India, the indigenous and naturalised silkworm races, and silkworm breeds evolved in India in different regions.

Among the evolved races in India, there are several mutants and sex-limited races of popular breeds in current use; which were mainly evolved for egg production purpose. By utilising these germplasm, the silkworm breeders in India have evolved high yielding silkworm races capable of yielding one kg of silk from 6 kg of cocoons. Also, some of the newly evolved bivoltine silkworm races and CSR hybrids have several desirable attributes like long silk filament length (more than 1200m) and high survival rate under tropical climate.

The silkworm germplasm maintained in the field gene bank at CSGRC, Hosur, in India exhibit wide variations for most of the heritable characters and therefore the germplasm collection is a valuable promising raw material for further crop improvement in sericulture.

High genetic variability in economic characters is a resource for breeding programme. The evolved breeds, land races, mutants, sex limited races, breeders genetic stocks and geographical races constitute the serigenetic resources in India, which are maintained in the germplasm centre at Hosur in the ex-situ field gene bank. Apart from these genetic resources, there are wild relatives of Bombyx, *B.mandarina, Ocinara* sp., *Theophila* sp. etc. and these constitute the wild serigenetic resources of India. However, they are not maintained in ex-situ field gene bank.

Realising the pivotal role of sericultural germplasm resources for sustainable development of sericulture industry in India, the Government of India (GOI)/Central Silk Board (CSB) established Central Sericultural Germplasm Resources Centre (CSGRC) at Hosur in 1991 to act as a trustee of sericultural germplasm for easy accessions by the breeders and other users for their research programme to utilise these valuable germplasm towards mulberry and silkworm crop improvement. The CSGRC, Hosur is mandated to function as the nodal centre for planning and coordinating all activities relating to sericultural germplasm viz. collection, characterisation, evaluation, conservation, supply and exchange of mulberry and silkworm genetic resources in India.

The centre has established a National silkworm gene bank for conservation and utilisation of the genetic resources. Since 1991, the centre has collected mulberry and silkworm germplasm readily available in various sericultural research institutes and universities and also carried out exploration and survey mission particularly in the Himalayan ranges and Andaman Islands. As a result of concerted efforts, the centre has collected and conserved 357 silkworm germplasm resources of diverse origin in the field gene bank. The silkworm germplasm conserved at CSGRC, Hosur are maintained with an unique accession number allotted by the Centre for easy identification of the germplasm. The centre has also developed a system for registration of sericultural germplasm to protect the breeders rights/Intellectual Property Rights (I.P.R.).

The germplasm centre maintains nearly 63 multivoltine and 294 bivoltine silkworm accessions. The multivoltine silkworms are maintained in a 5 rearing cycle per year and the eggs are preserved at low temperature (5°C) for 30 days, whereas the bivoltine silkworm accessions are multiplied once in a year during favourable season, by consigning the eggs in 10 months preservation schedule at low temperature regimen. In order to maintain the genetic variability within the population, composite layings are prepared from 60 layings so that all the heritable variability within the population are maintained, since the genetic variation of the base population is very essential for genetic improvement in silkworm race evolution.

Since the germplasm is a valuable genetic material, it is very essential to maintain the accessions as per the genetic characters listed in the passport data and preserve the genetic variability and the integrity of the race/breed. For this purpose, 60 layings are prepared; out of which 40 layings are preserved in one cold storage and the remaining 20 layings are preserved in another cold storage as a safety back-up. The 40 layings are further divided into two equal parts and consigned at two different schedules; and one set is kept as a buffer stock. From one set of the 40 layings two composite eggs are prepared at pin head stage of the egg and two batches of rearing is conducted; simulating two replications.

In the rearing process, all the worms are maintained upto third instar and thereafter only 300 healthy worms are retained in multivoltine and 250 worms in bivoltine. From each replication, 200 to 250 healthy cocoons are selected for egg production for continuing the next rearing cycle and the details are given in fig-1 & 2 for multivoltine and bivoltine respectively. During the multiplication cycle adequate care is taken to maintain the original characters of the genotypes, as per the passport data.

The multivoltine accessions are reared five times in a year and the eggs are preserved in the cold storage for 30 days at 5°C. One rearing cycle of multivoltine accessions takes 75 - 80 days including egg preservation. The bivoltine accessions are divided into three batches with nearly 100 accessions in each batch and grouped as I, II & III batch respectively and the rearing of these three batches are conducted in Jan-Feb, July-Aug and Aug-Sept. respectively, and the eggs are preserved under 10 months schedule in the cold storage. One rearing cycle of bivoltine accessions takes 360 - 370 days.

In India, sericulture is practiced in widely varied agro climatic conditions like hot and dry tropical climate, hot and humid tropical condition, sub tropical conditions and temperate climatic conditions and also under different cultivation practices viz. rainfed and irrigated, which require different silkworm races with desirable genetic potentials. In hot and dry as well as hot and humid climatic conditions in the tropical regions, bivoltine and univoltine silkworms do not survive and hence multivoltine silkworms and cross breeds (multivoltine female × bivoltine female) are reared in the tropical regions.

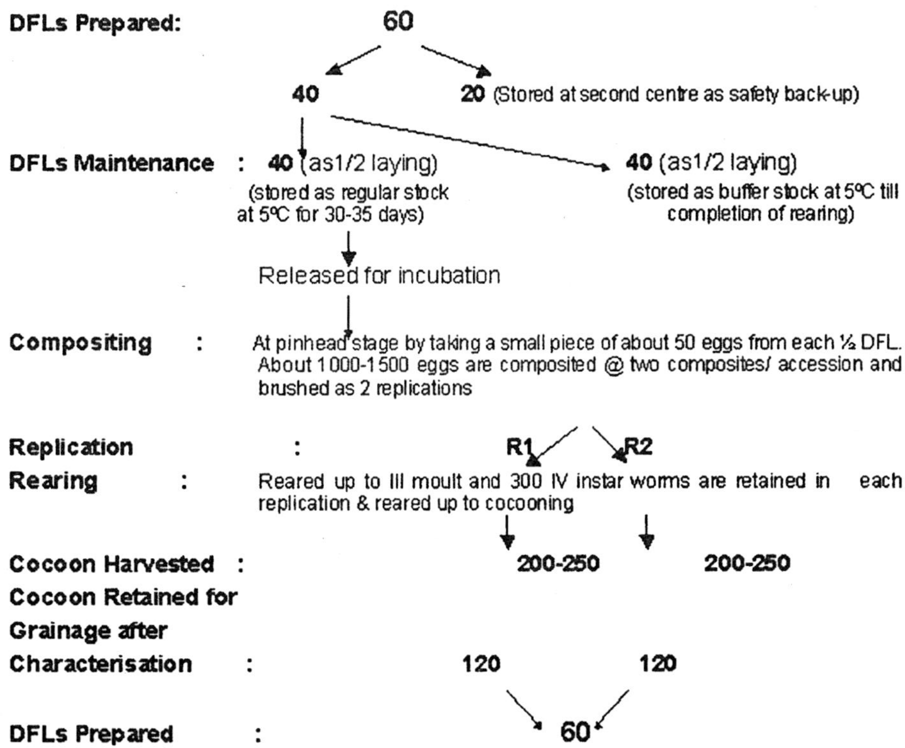

Figure 1. Plan of conservation for multivoltine silkworm germplasm

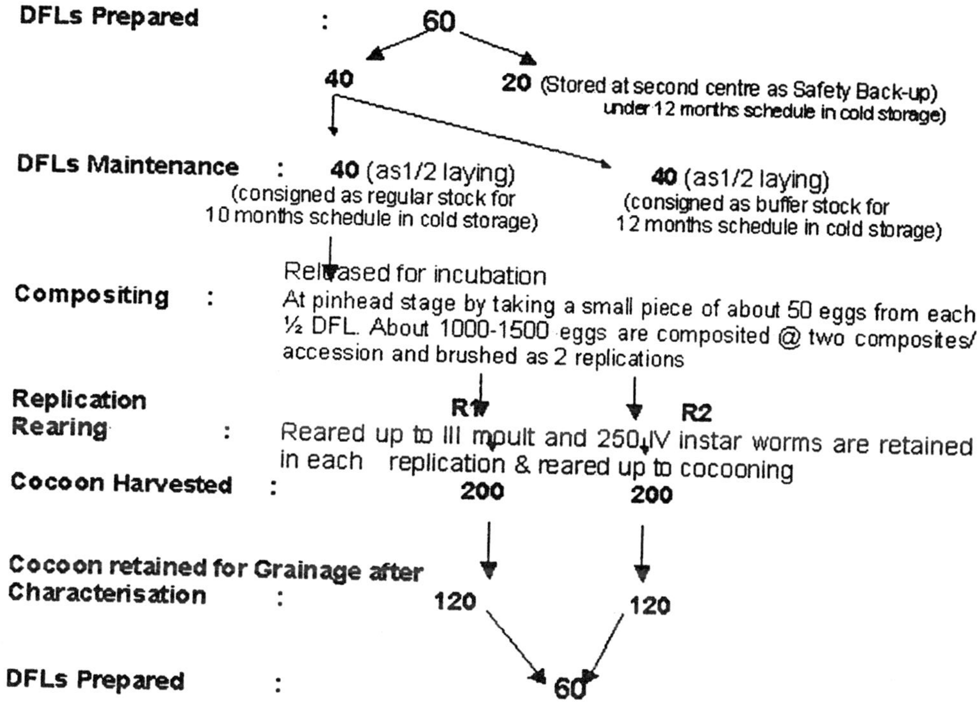

Figure 2. Plan of conservation for bivoltine silkworm germplasm

Whereas in the subtropical and temperate regions, bivoltine hybrids are reared during the favourable seasons and hence India requires multivoltine as well as bivoltine silkworm races to suit different climatic conditions. For rainfed condition, hardy silkworm races are required. Thus, the varied agroclimatic conditions and sericultural practices prevailing in different parts of India require 'season' and 'region' specific races / breeds / hybrids / cross breeds etc. Therefore, several silkworm breeding centres are established in different agro-climatic regions to evolve silkworm breeds suitable to the specific region and sericultural practice.

Hence, the silkworm germplasm are maintained in a networking mode in different centres, representing different agro-climatic zones. This arrangement for silkworm conservation serves as a safety back-up and also facilitates silkworm breeding. The main centre at Hosur is the base collection and national repository and the regional centres are active collection centres. This arrangement is working extremely well and this also facilitates characterisation and evaluation of silkworm germplasm in the different agro-climatic condition and the data provide basic information on Gene × Environment interaction.

Since silkworms are highly susceptible to diseases and abiotic stresses, the multiple conservation under networking plan ensures total safety of the silkworm germplasm and maintaining multiple populations enhances the chances of their long term survival, in terms of conservation. The silkworm germplasm conservation strategy adopted in India is a fail-safe device and ensures total safety of the silkworm germplasm and also promotes greater utilisation of the silkworm germplasm for breeding and other studies.

In addition to the existing field gene bank in the National repository in the main centre and active germplasm collections in the net working centres it is proposed to establish DNA bank for silkworm, so that total genomic DNA can be preserved for utilisation to develop 'transgenic silkworm' and other molecular studies in *B. mori*. Similarly, the wild relatives of *B. mori* and species of *Ocinara* and *Theophila* belonging to the family *Bombycidae* are proposed to be conserved under *in situ* conditions, so that there could be gene-flow from the wild sericigenous insects to the *Bombyx mori* for widening its genetic diversity.

Silkworm Genetic Resources Management

For sustainable development of sericulture and progressive increase in productivity and improvement in raw silk quality, we require continuous evolution of highly productive breeds of silkworm with desirable traits viz. short larval duration, high survival rate, higher fecundity, higher silk ratio, low denier, long silk filament length, higher reelability, low boil off loss, higher tenacity and cohesion etc. Also, silkworm races tolerant to adverse climatic conditions and resistant to diseases are required. To meet these challenges effective utilisation of the available genetic diversity in silkworm is the only answer.

The silkworm germplasm available in Japan, China, India, Russia, South Korea, France and Italy constitute a great resource of silkworm breeding material, other countries like Brazil, Iran, Thailand, Indonesia, Vietnam, Bulgaria and Egypt also maintain some valuable silkworm germplasm. These silkworm germplasm exhibit wide genetic variations and therefore form a nucleus for evolution of silkworm breeds with specific characters viz. short larval duration breed, breeds tolerant to diseases, breeds with long silk filament length, breeds with high survival rate, breeds with high pupation rate, breeds with low denier silk filament etc.

Agricultural crop improvement witnessed great progress after establishment of International Board for Plant Genetic Resources (IBPGR) in 1974 and the International Agricultural Research Centres (IARC) assembled diverse gene pool of wild germplasm along with cultivars, land races, breeders genetic stocks and supplied them to the needy countries for crop improvement programme under bilateral agreement mediated through Food and Agricultural Organisation (FAO). Whereas, in sericulture crop improvement programme, such International Co-operation and periodical/need based exchange of silkworm germplasm is not taking place between countries due to various barriers and silkworm germplasm are still considered as exclusive possession of the nations.

Whereas, international co-operation in sericulture is very much limited, particularly, Japan has liberally assisted some countries like Brazil, Iran, India, Thailand, Indonesia and few other countries through Japan International Co-operation Agency (JICA) for development of sericulture in these countries, mainly with financial assistance, training in sericulture, and to a limited extent in silkworm breeding. However, exchange and supply of silkworm germplasm from Japan is very much limited.

Similarly, South Korea, China and India extend training facilities in sericulture on specific request from developing countries and to a limited extent the silkworm germplasm have also been exchanged under bilateral agreements or mutual co-operation. For instance, India extended technical co-operation in sericulture to Sri

Lanka under Indo-Colombo Plan and South Asian Association for Regional Co-operation (SAARC).

The Govt. of Switzerland has funded International Sericulture Training facility in tropical sericulture to developing countries through Swiss Development Co-operations (SDC) and the programme is conducted at Central Sericulture Research and Training Institute (CSR&TI), Mysore under the auspices of Central Silk board (CSB), Govt. of India (GOI) and some of the developing countries have received sericultural germplasm from India, apart from training in sericulture. However, these exchange of sericultural germplasm (Mulberry and Silkworm) at the global level among the countries is very much restricted, at time through diplomatic channel or personal contacts, and there is no need based and regular exchange of silkworm germplasm among the countries for silkworm breeding and other research purpose, as witnessed in the case of agriculture, horticulture, animal husbandry etc.

In fact, great progress has been made in food production particularly in wheat, rice, maize etc. in the developing countries, mainly because of the regular exchange of crop germplasm and agro-biodiversity through IARCs, viz., International Rice Research Institute (IRRI), Philippines, Centro International de Mejoramiento de Maiz Y Trigo, (CIMMYT), Elbatan, Mexico.

The very purpose of germplasm collection and conservation is for utilisation to benefit the human society. The silkworm germplasm is the basic raw material for improving the quality of silk as well as increasing the productivity of silk; for this purpose elite and robust silkworm breeds resistant to diseases and tolerant to adverse climatic conditions are required, which can come mainly from silkworm breeding. The silkworm germplasm are therefore utilised to evolve breeds, rejuvenate the popular breeds and identification of potential donor from the available genetic resources through evaluation and selection.

The CSGRC, Hosur is currently maintaining 357 silkworm germplasm (63 multivoltine and 294 bivoltine silkworm accessions). The silkworm germplasm are supplied to Research Institutes and Universities for crop improvement programme and the supply is regulated through Germplasm Supply Committee constituted for this purpose. Some of the silkworm germplasm are approved for supply to Research Institutes based on their indents submitted in prescribed format and the Director, CSGRC is authorised to supply those authorised silkworm germplasm.

Indents which are outside this authorised silkworm accessions are screened out by the Germplasm Supply Committee and based on the merits of the indents and specific study, the silkworm germplasm supply is made. The request from other countries for supply of silkworm germplasm are decided by the Member Secretary, Central Silk Board, Govt. of India on merit basis. Thus, during the last seven years CSGRC, Hosur has supplied 445 silkworm collections to 45 indentors.

At the time of collection, the silkworm germplasm are collected with a passport data and also examined for diseases, pest etc. and if any pathogens or pest are identified, such materials are rejected and not transported to the germplasm centre. In the case of silkworm, the eggs obtained from the collection site are disinfected in the quarantine laboratory by way of surface sterilisation. This is the preliminary quarantine. Germplasm materials that are free from pest and diseases only are transported to the germplasm centre.

On arrival at the germplasm centre, the silkworm accessions are first taken to the introduction laboratory, which is isolated, and the rearings are conducted carefully with periodical examination for disease at every stage, if any diseases are identified, such materials are screened out and rejected; this is the next stage of the quarantine and only those which are free from pathogens and disease or pest, such accessions transferred to the base collection. Thus, the silkworm germplasm accessions are collected with careful methodology to exclude exotic pests and diseases from entry into the germplasm centre.

Similarly, while supplying the silkworm accessions, the germplasm materials are thoroughly subjected to examination of pests and diseases and only those materials that are free from pests and pathogens are supplied along with quarantine certificate in case of silkworm. By this method, the introduction and spread of pests and pathogens of mulberry and silkworm are strictly avoided, and International standards of Phytosanitory and Quarantine measures are implemented at the time of collection, introduction and supply of silkworm germplasm.

Germplasm Database

For sustainable and regular use of germplasm, the foremost requirement is the information about the germplasm. Therefore, characterisation, documentation and information management are very essential aspects of germplasm management. During the course of germplasm collection, characterisation and evaluation, considerable information on the germplasm are generated. Germplasm maintenance and conservation are useful only when they are utilised effectively on a continuous basis. However, for wider and regular use of germplasm, information/database are essential.

Therefore effective utilisation of germplasm resources mainly depends on the information documented so that the breeders and other researchers can objectively select the desirable genotype required for specific breeding and other research programmes. The CSGRC, Hosur has developed an information package for silkworm germplasm data maintenance and retrieval, Silkworm Germplasm Information System (SGIS). The information package facilitates data maintenance, analysis, data listing and easy retrieval with facility to multiple query. The germplasm information system maintains passport data, collection data, characterisation data, evaluation data and conservation information.

The germplasm information management system developed at the Hosur centre is flexible and facilitates easy data entry, data evaluation during input stage and multiple analysis and therefore the information management system is user friendly. The SGIS was developed using integrated data base technique incorporating user friendly menu driven facilities for easy data view, information retrieval, updation and report generation. Each file of the integrated data base is linked to the accession number of the germplasm which is an unique number assigned to each silkworm germplasm accession serving as the primary key.

The systems are flexible enough to allow modifications and insertions of new modules as per the changing needs. Also, the facility for identification of the accession(s), on the basis of certain desirable criteria up to 10 quantitative and/or qualitative characters has also been developed as separate module. This information system promotes utilisation of germplasm in large scale. Utilising the SGIS, the first volume of catalogue on silkworm germplasm was compiled and published for circulation among the breeders and other research workers.

Descriptor/stock characterisation

To promote utilisation of silkworm genetic resources, the following activities are undertaken at the germplasm resources centre.

1) Characterisation of silkworm germplasm and data base generation.
2) Evaluation of silkworm germplasm for biotic and abiotic stresses in the specific hot-spots
3) Evaluation of silkworm germplasm for yield potentials at multi-locations.

Silkworm races from different geographical regions exhibit considerable variations for several important heritable characters viz., egg and cocoon colour, shape and size, voltinism, larval markings, moulting behaviour, amylase activity in the digestive juice, carotenoid and flavonoid content in the silk fibre etc., Similarly, there are distinct molecular differentiation at the genome level in different silkworm accessions. Therefore, a proper and exhaustive characterisation will help to identify the distinctness of different silkworm accessions through phenotypic, biochemical and molecular markers; which will help selection of parents for silkworm breeding.

Through characterisation, the specific identity of each accession is maintained, which helps to distinguish the accessions. Characterisation is mainly based on the heritable quantitative and qualitative characters, viz. morphological, anatomical, cytological, embryological, physiological, biochemical and biomolecular aspects. Characterisation also facilitates registration of germplasm and thereby establishes the Intellectual Property Rights (IPR).

In the case of silkworm germplasm characterisation a set of descriptors containing 'collection data', 'passport data', 'Specimen data', 'characterisation descriptor', and 'evaluation descriptor' are utilised for generation of information on the silkworm germplasm. The collection data elicit information on origin, location, geographical details of the accession. The passport data contain information on source and history of the accession.

The specimen data contain bio-data of the accession, including parentage, voltinism, larval, cocoon and silk characteristics. These informations are obtained at the time of collection along with the accession. At the time of introduction of the accession into the base collection, accession(s) are studied meticulously for 27 characters (descriptor - characterisation) of egg, larva, cocoon, pupae and adult.

Molecular Characterisation

Systematic genetic analysis of quantitative and resistance traits in silkworm has not been done, although the correlated changes in different quantitative traits are well understood Quantitative Trait Loci (QTL) mapping has become very powerful tool in plant and animal improvement programme and silkworms offer as a very interesting insect for such exercise. There is no comprehensive method available for genetic characterisation of the silkworm gene pool at the molecular level.

However, the well known molecular markers viz., Restriction Fragment Length Polymorphic (RFLP) markers, Minisatellite DNA markers, Microsatellite DNA markers, Random Amplification of Polymorphic DNA (RAPD) markers, Inter Simple Sequence Repeat (ISSR) markers and Amplified Fragment Length Polymorphic

(AFLP) markers provide valuable information to identify duplicate accessions apart from trait based mapping and molecular tags.

Molecular characterisation helps to identify the genetic distinctness of a race/breed/stock and thereby helps to eliminate duplicates and reduce the cost of germplasm maintenance and volume of work. Thus, the genetic markers facilitate molecular characterisation, maintenance, conservation and cost effective management of plant and animal genetic resources. Molecular markers that are linked to metric and resistance traits and their subsequent mapping in silkworm is essential. Also these genetic markers can be applied periodically to monitor changes in heterogeneity and heterozygosity in the accessions, as they are routinely reproduced for maintenance.

These molecular markers provide essential information for the development of "Core Collections". Molecular markers also facilitate genome specific and accurate information on genetic variability that may escape during characterisation of a race based on morphological, anatomical and biochemical parameters. Such molecular characterisation promote molecular marker assisted selection which will eliminate narrow genetic bases, inbreeding, common pedigree etc.

Molecular characterisation of 56 multivoltine and 80 bivoltine silkworm accessions were carried out; using SSR-Anchored PCR method involving amplification of the regions between pairs of microsatellites and anchored repeat primers to generate multiple polymorphic bonds. Inter-microsatellite DNA polymorphism was also studied by using two anchored primers. The primer [(GCT) (AGT) (GCT) (CA) TC] amplified a high number of polymorphic bonds.

The polymorphism is quite evident among the accessions tested. PCR amplification of [(GT) 8 (A/G) TCC] primer also amplified large number of bonds, which shows less polymorphism among the accessions screened. The anchored primer identified a high cocoon weight marker of approximately 182-183 bp size mostly in high yielding silkworm accessions. Different statistical analysis were carried out to make use of the DNA profiles generated by the two primers. The discriminant function analysis to test the validity of the zone identification with regard to polymorphic DNA profiles indicated five discrete groups.

The hierarchical clustering analysis based on yield parameters indicate that some of the accessions are very unique. The results indicate that the silkworm accessions available in the Indian germplasm collection are from different zones and these biotechnological tools can be used for analysing the actual genetic differences/genetic distance among the accessions based on the genetic characters which are more reelable, rather than the phenotypic characters alone.

Evaluation of Germplasm

Evaluation of germplasm promotes effective and greater utilisation of the germplasm, particularly in breeding and crop improvement. At the time of introduction of the accession into the base collection 38 characters of evaluation parameters are studied. The evaluation descriptors are mainly for commercial characters viz., cocoon yield and raw silk quality.

Evaluation of germplasm under abiotic and biotic stress conditions is very essential to identify promising germplasm for further utilisation in specific crop improvement programmes. The promising germplasm are further tested under multilocational trial to assess the genetic potential and genotype × environment interaction and genotype stability. The silkworm accessions undergo preliminary evaluation at CSGRC, Hosur for identification of the genetic potential of the accession and the data are utilised for classifying them as high yielding / tolerant to high temperature / disease tolerant / short larval accession etc., based on their performance in the preliminary evaluation trial.

In the preliminary evaluation the accessions are subjected to uniform treatment and their response are evaluated from the performance. The preliminary evaluation trial is conducted for 3 crops to study the potential and stability of the characters. The silkworm breeds are also subjected for evaluation under different seasons. The silkworm races are evaluated for 13 rearing parameters and 16 silk characters. From the preliminary evaluation data, the silkworms are classified and grouped based on Warts' minimum variance method to identify the genetic relationship between the clusters and this method helps in grouping the silkworm germplasm. Such study will help the breeders to select the silkworm germplasm required for specific breeding studies.

After screening the silkworm germplasm through preliminary evaluation, the potent / selected accessions are subjected to screening against various biotic and abiotic stresses to identify tolerant / resistant breeds under hot spots / sick beds. The biotic stress include various pathogens (protozoan, bacterial, viral and fungal). The results of such hot spots / sick beds evaluation indicate the most potential accessions / breeds which are tolerant / resistant to specific biotic stress. Similarly the selected silkworm germplasm after preliminary evaluation are subjected to abiotic stress, particularly the adverse climatic condition (high temperature and high humidity / high temperature and low humidity).

Preliminary evaluation facilitates identification of silkworm clusters for resistant to a specific biotic or abiotic stress and only the selected accessions are subjected to sick-bed/hot spot evaluation for further estimation of the genetic potentials of the selected breeds and the best performed breed is used directly in the breeding programme. Hot spot evaluation of silkworm to important biotic and abiotic stress is considered as most useful evaluation to identify the silkworm breed with useful genes conferring better adaptation to stress environment and resistance to disease and pests.

The hot spot evaluation is conducted in different regions with collaboration of the networking centres. The screening of silkworm germplasm for biotic and abiotic stresses in the in-situ conditions under hot spots/sick beds provides objective assessment and evaluation of the germplasm and the information generated helps the silkworm breeders to select the right parents for specific silkworm breeding programme.

All India Mulberry and Silkworm Germplasm Evaluation Programme (AIMSGEP)

The CSGRC at Hosur has identified some silkworm races as best performing by multiple scoring analysis and these accessions are either at par or even performing better than the popular silkworm breeds commercially exploited in the field. In order to select the best performing breeds, the identified silkworm breeds are tested in different biogeographic zones along with local and national checks (popular breeds). The first field trial of 10 silkworm races is under trial at different places. The consistently best performing silkworm races will be sent for race authorisation trial, before release to the field in the particular area where the race performs better. The race authorisation trial is a multi-location trial to identify region/season specific breed.

Promotion of Global Sharing of Germplasm Information

The information sharing on silkworm germplasm at global level among the countries is very much limited, unlike in agriculture, horticulture, sericulture, animal husbandry and fisheries etc. Hence it is very essential for sharing the information on silkworm germplasm among the nations, which will promote silkworm breeding and sericulture development. In several countries sericulture is not sustainable and cost of production of silk is very high mainly because suitable silkworm breeds are not available and they are also unable to evolve silkworm breeds suitable to their own agro-climatic conditions. Therefore, it is essential to share the information on silkworm germplasm among the countries involved in sericulture.

Standardisation of formats

The terminology(ies), evaluation, estimation methods and the information formats used for descriptor characterisation and evaluation need standardisation to promote international exchange of silkworm germplasm, since different countries may be adopting their own methods / standards / terminology / formats etc. Also a set of core characters may be identified for uniform characterisation in all the silk producing countries; based on which clustering and core collection of silkworm germplasm can be made; core collection will actually facilitate the silkworm breeders in decision making to identify the parents for breeding and also select tester and lines for hybridisation from among the elite accessions.

Publication in common language and circulation

Sericulture is practiced in different countries across Asia, Africa and Latin America and the languages spoken in these countries are quite different and therefore there is a communication barrier among the nations. This is an important bottleneck for sharing the information on sericulture germplasm and sericulture related activities. Therefore to facilitate sharing of information on sericultural germplasm, it is essential to publish important information on sericulture germplasm in common language for circulation among the nations.

Internet uploading, sharing and establishment of internet links

The internet is a "network of networks" and operates through the computer systems around the world using a common set of rules called 'Transmission Control Protocol' or 'Internet Protocol'. It is very useful for speedy exchange of information. Using the internet facility many organisations around the world have developed their own Website and display their products, information and promote their business. The internet facility helps scientific research also for exchange of information. On the same lines, internet facility may be used for the development of website containing data on sericultural germplasm to facilitate exchange of information and promote utilisation of sericultural germplasm.

Global exchange of silkworm germplasm

Sericulture is practiced in 53 developing and under developed countries benefiting 150 million people mainly from rural area and economically weaker sections of the society. Therefore, sericulture development will improve the quality of life of economically poor people in under developed and developing countries, for which high yielding mulberry varieties and robust silkworm breeds, and breeds tolerant to adverse agro-climatic conditions are essential; which means the available mulberry and silkworm germplasm need to be exchanged among the countries for the benefit of mankind in the world.

FAO may promote silkworm breeding among the member countries involved in sericulture and facilitate need based exchange of sericultural germplasm among the member countries. The silkworm and mulberry germplasm available in Japan, South Korea, France, Italy, erstwhile United Socialist Soviet Russia (USSR) and other countries where sericulture has declined may be deposited in China/India for conservation and utilisation among the FAO member countries, for which FAO may provide financial assistance and infrastructure facilities, similar to the International Agricultural Research Centres functioning under the auspices of Consultative Group on International Agricultural Research (CGIAR) and facilitate exchange of sericultural germplasm similar to International Board for Plant Genetic Resources (IBPGR), Rome.

To organise an international centre for conservation of sericultural germplasm and formulate the procedure, terms and conditions for the exchange of germplasm among the FAO member countries, the FAO may constitute a Working Group and prepare a proposal for sanction by FAO.

Germplasm Resources in Japan

It has been inferred that the utilisation of cocoon began in China 6000 or 7000 years ago. Ancient Chinese characters used before 11th century BC, such as thread and mulberry, show the existence of the sericulture. The sericulture had spread from China to the world. The naturalisation person who came from each of Korea and China to Japan brought silkworm and technology of the textile between BC 300 and AD 300. It is uncertain for whether the early races of Japanese silkworm are improved as genetically pure race or not. Though the ancient description describing the existence of Japanese silkworm race have been left in great numbers since 1700s, the most varieties were not pure, and the quality was not stabilised.

Japan solved the national isolation in 1859, and the trade was started by opening of Yokohama port. The sericultural industry became the important industry of Japan for acquisition of foreign currencies after that. However, there were many complaints about the irregular quality of Japanese silkworm race especially from USA where was the main destination for export in those days. So the Japanese government prohibited farmers the rearing of non-permitted races between 1937 and 1998. At the present, rearing and distribution of any silkworm have been permitted.

Imported silkworm races had an important role for Japanese sericulture. Since 1859, many silkworm races were imported from foreign countries, and many races were bred in Japan. The newly bred races were mainly contributed to establish pure lines to produce F1 hybrid. Subsequently, main races for cocoon production were shifted to hybrid races at the beginning of 20th century.

In 19th century or earlier, the most of cocoons produced in Japan were mainly consumed domestically rather than its export. At the era, major race of domesticated silkworm was Japanese old varieties. Since the government encouraged the sericulture in 1870, the cocoon production increased. At the same time, major silkworm race was shifted to the hybrid races between

Japanese race and the pure race bred from imported races. In 1920's the cocoon production reached the record, more over 350 000 ton per year. However, the export decreased sharply at the beginning of the Pacific War in 1941. After the war cocoon production gradually declined and finally decreased to 1 300 tons.

To prevent the decline of Japanese sericulture, new silkworm races those had value added to obtain the largest profit was bred. By 1800's, silkworm breeding was carried out as autonomous activities of the farmers. In Japan, sericulture region had been divided nationwide into two purposes, for the cocoon production and the egg production. From 1859, at the beginning of raw silk export, native Japanese silkworm races were used for sericulture without newly breeding. Since 1937, the government prohibited the rearing of non-approval races except for the race that had high and uniform quality to produce raw silk.

By the establishment of the license system for the silkworm-egg production and distribution, farmers were forbidden to breed the silkworm freely, and the most of the breeding activity had shifted to experimental stations and private companies. At that time, the breeding goal was focused on the high yielding and healthiness to enhance the further production of standardised raw silk. Since 1970s, however, the cocoon production of Japan was reduced by the pressure of increased raw-silk import, so the income from cocoon production had become too little to support farmer's life. The change of trade balance caused silkworm breeders to change the breeding goal to characteristic silkworm race.

In the modern period, from 1800s to 1900s, many silkworm races were imported from Italy, France, Korea and China. Thereafter, many races were bred from the hybrids between Japanese native race and imported races. Recently, as the number of silkworm races has been more than 500, import of silkworm races is not necessary only by using such races. As an example for recent international activity by Japanese researchers, technical cooperation of the sericulture by bivoltine race is made between Japan and India to promote new sericulture in India where the polyvoltine race sericulture is carried out.

Recently, the breeding activity of domesticated silkworm has been performed especially in National Institute of Agrobiological Science (NIAS) in Japan. In NIAS, the breeding for the production of the standardised raw silk and additional-value race are going on. The breeding goals are e.g. thicker filament race, thinner filament race, oligophagous race and virus-resistant race. On the other hand, the breeding of silkworm races adapted to factory production of physiological-active substances by the use of the vaculovirus-vecter system are proceeding in NIAS. The races are expected to make the great industry in Japan. Since Tamura succeeded to make transgenic silkworm in NIAS, new silkworm strains has been increased.

Conservation

In Japan, silkworm races are consists of native race, improved race and mutants. The silkworm races owned by NIAS are grouped into three major categories: silkworm genetic stock, breeding resources and currently popularised race. The later two categories were excluded from the genetic stock because their characters are varied by annual selection depending on the change of public needs, and the distribution of the eggs are left for the breeder to keep their title to an race. Such races will become prospective the genetic stock.

In the case of mutants, the largest stock is conserved in Kyushu University while NIAS stock is only 172 lines. By the way, the most popularised race has been bred and distributed by private companies. Additionally, Gunma Prifecutural Experimantal Station of Sericulture and some other experimental stations own original races for districted sericulture too. The genetic stocks of NIAS are mentioned in the following paragraphs.

Present Status

Six hundreds and forty five lines of domesticated silkworm have been preserved as the genetic stock of NIAS. The genetic stock is greatly classified into four types: native race, improved race, mutant and intermediate stock. The native races are classified by geographical origin, like the tropics, Japanese, China and Europe. The improved variety is classified into three categories, Japanese, Chinese and European races. Although, the detailed description is omitted, mutants are preserved mainly for the experimental purpose.

The breeding middle mother plant has not been completed as a variety, but it recognised that the line is a characteristic compared with common commercial races. The breeding section of NIAS possess many lines, those are not yet registered to the genetic stock. It could be guessed that the number of such lines reach 50 - 200 per laboratory. Those lines are improved and newly established in each rearing period, so we cannot count the accurate number.

NIAS is managed by the budget from Ministry of Agriculture, Forestry and Fisheries and expected to perform the contribution to the Japanese agriculture. NIAS established Gene Bank Project in relation to many genetic resources: crops, domestic animals, insects including silkworms. In the project, great number of the genetic stock has been preserved and distributed. The half of the budget necessary for the maintenance of the genetic stock has been supplied from the gene bank project. It manages the remainder by the research budget of NIAS.

Previously, over 2000 lines of silkworm were maintained in Japan, and 1000 lines of the genetic stock had been preserved in Sericultural Experimental Station, it's the former of NIAS. However, the preservation system is reduced with the decrease of farmers engaging to sericulture, consequently the number of the genetic stock had decreased. Recently, the importance of the genetic stock was well understood by the nation, so the decrease in the number has been stopped. On the contrary, present NIAS became possible to accept the additional preservation of silkworm germplasm. As the gene-bank project have the plane to evaluate its outcome in every five years, the person, who have the charge to preserve the genetic stock, has to make efforts to obtain the budget in each period.

Management and Utilisation

Except for the breeding resources now on use, the genetic stock is preserved in the laboratory of insect genetics in NIAS. The breeding resources have been preserved in Laboratory of Sericultural Science and Laboratory of Insect Molecular Evolution. The present state in the laboratory of insect genetics is described in the following paragraphs.

Methods for Maintenance

The silkworm races were bred to make homosygote for important characters in relation to cocoon filament and to maintain heterozygosity in other characters, like disease tolerance. It is concerned that inbreeding makes the silkworm race wrong. Therefore the silkworm breeders always select the excellent individuals or lines from the silkworm race in each generation to maintain the average performance owned by itself. It has been suggested historically that the maintainance of excellent character of race is difficult because of less number of silkworm race reared continuously more than ten years in Japan.

The laboratory of insect genetics is preserving 470 lines of the genetic stock by the hands of five staffs and seven part-time workers. The number of persons is too small to perform the selection in all races. Therefore, the laboratory gave up a part of selection step like weighing the individual cocoon and reeling test. In other words, the laboratory abandoned the definition that silkworm race always express its own phenotypes, while concerned that a race is the pool of the excellent genes. If we are able to avoid the loss of excellent genes without the selection in a race on every generation, the race will revived again by re-selection in future.

The pure lines, except for the tropical races, are reared one time per year and preserved in the stage of diapause egg by next year. Tropical races are reared at least two times per year because of the difficulty to obtain diapause egg in Japan and to avoid long-term chilling that make the hatchability to go down. Three rearing periods has planned to preserve the genetic stock: The first, second and third period are May-June, July-August and September-October respectively. The genetic stock including 470 lines has been divided into three groups reared in different period each other.

The diet for silkworm is mulberry leaf in all cases in order to avoid the unconscious selection by artificial diet. Since the allele of the low frequency would not be lost by the genetic drift, the egg is collected from fifty of female moth even in the minimum for the mating. Then, the larva is derived form all moth batches and supplied in the rearing in the next generation. In third inster, 160 of middle-sized larvae are usually selected from each race. After cocooning, 120 of cocoon that looks like good shape are selected to avoid the accidentally mixing of other races.

The cocoons were cut, and pupae are moved into the mating box before the emergence of moth. Consequently, approximately 50 egg batches are obtained for the preservation. The individual selection for the preservation is milder than the case of improving, and is performed simply by the removing of the individual that has incorrect character. Usually, hybrid F1 is not produced in the laboratory, because of no public demand about old hybrid. If current race is needed, the F1 hybrid is produced in the specialist section of NIAS. The pure lines of popularised hybrid are mainly preserved in Laboratory of Sericultural Science.

Various characteristics are investigated whether the race is surely preserved through whole preservation process of the genetic stock. In Gene-Bank Project of NIAS, It is imposed primarily that characteristic data of each race are registered. Character items of the gene bank are divided into three groups: primary character (Table 1a), secondary character (quantitative character [Table 1b]) and tertiary character (practical and economical character [Table 1c]). At least the clarify of the primary character is necessary to register silkworm race to the gene bank. Moreover, the primary character is reconfirmed in every generation.

Though the secondary character is dispensable to the registration, it is always recorded in the preservation. The

Table 1a. Primary characters needed to authorize silkworm race as genetic stock of NIAS

Check item	*Example of characters*	*Method or comment*
Voltinism	Univoltine, bivoltine or polyvoltine	Number of eggs laid
Egg color of diapause egg	Purplish gray, greenish gray, gray, pinkish gray or others	After overwintering
Chorion color	White, muddy white, yellow and light yellow	After hatching
Moltinism	Trimolter, tetramolter, pentamolter	During larval stage
Body color of newly hatched larva	Dark brown and reddish brown	After hatching
Body color of last inster larva	Bluish white, yellowish white, reddish white, translucent, black, yellowish orange or moricaud	Last inster larva most gluttonous
Body shape of last inster larva	Thinner & longer, normal, thicker & shorter, bigger and smaller	Last inster larva most gluttonous
Larval body marking	Normal marked, pale marked, plain or mixed	Last inster larva most gluttonous
Abdominal larval leg color	White and yellow	Last inster larva most gluttonous
Cocoon color	White, green, light green, greenish yellow, yellow, golden yellow, vermilion, flesh-colored or others	After harvesting of cocoon
Cocoon shape	Fully narrowed bale, slightly narrowed bale, ellipse, longer ellipse, spherical, spindle, cotton like, sericine cocoon or others	After harvesting of cocoon
Cocoon wrinkle character	Normal, dense, rough and flossy	After harvesting of cocoon Pupal
	Normal, black pupa, black wing or naked pupa	After harvesting of cocoon
Marker gene	Anything known	Marker gene of egg, larva, pupa and cocoon should be described.

Table 1b. Secondary characters needed to authorize silkworm race as genetic resource of NIAS

Check item	*Example of characters*	*Example of characters*	*Method or comment*
Hatchability versus total fertilized egg	Percentage (round off to one decimal places)	Percentage	Average of numbers counted
Number of eggs laid	Number (round off to two decimal places)	Number	Average of numbers counted from 5 egg batches
Duration of last larval instar	Number of days (round off to three decimal places)	Number of days	From first feeding of last instar larvae to mounting
Duration of larval stage	Number of days (round off to three decimal places)	Number of days	From brushing to mounting
Percentage of missing larva	Percentage (round off to one decimal places)	Percentage	=(Number of cocooning larvae/number of larvae at the beginning of 4th inster) × 100

Table 1c. Tertiary characters needed to authorize silkworm race as a genetic stock of NIAS

Check item	*Example of characters*	*Method or comment*
Cocoon weight	g (round off to three decimal places)	Average of values derived from 25 cocoons
Cocoon shell weight	cg (round off to two decimal places)	Average value derived from 25 cocoons collected separately between male and female
Percentage of cocoon-shell weight	% (round off to two decimal places)	= (cocoon shell weight/cocoon weight) × 100
Cocoon-shell lousiness	Point (round off to three decimal places)	Average value derived from 25 cocoons Standard photograph is compared to samples
Degumming loss of cocoon shell	% (round off to two decimal places)	Average value derived from 25 cocoons = (cocoon-shell weight - the weight after degumming)/cocoon-shell weight
Raw silk percentage	% (round off to three decimal places)	Average value derived from 50 cocoons =gross weight of raw silk/(fresh cocoon weight × number of cocoons used for reeling test
Length of cocoon filament	Number (integer)	Average value derived from 50 cocoons = (raw-silk length × average number of cocoon fed at the same time to button during reeling)/total number of cocoons used at one times
Weight of cocoon filament	cg (round off to two decimal places)	Average value derived from 50 cocoons = gross weight of raw silk/number of cocoons used in reeling test

Check item	*Example of characters*	*Method or comment*
Weight of cocoon filament	cg (round off to two decimal places)	Average value derived from 50 cocoons = gross weight of raw silk/number of cocoons used in reeling test
Size of cocoon filament	denier (round off to three decimal places)	Average value derived from 50 cocoons = (gross weight of raw silk × 9000)/(raw-silk length × average number of cocoon fed at the same time to button during reeling)
Reelability of cocoon	% (integer)	Average value derived from 50 cocoons = number of cocoons used in reeling test/number of times of feeding in reeling test
Neatness defects of raw silk	point (round off to two decimal places)	Standard photograph is compared to samples

tertiary character is investigated in the every generation of currently popularised race, while it is not always for the genetic stock. The investigation of the tertiary character needs rearing the double number of larvae at once. The increase of the number of larvae limits the number of the genetic stock to half because of facility. Therefore, the investigation should have been carried out at least one time per ten years in the preservation of the genetic stock. The last investigation was reported in 1986, and it has not been investigated afterwards.

Development of Silkworm Germplasm Resources

It is supposed that Japanese-native race was improved empirically by the pure-line selection segregated from hybridisation. In 1914, Dr Toyama introduced the usefulness of heterosis to Japanese people. Thereafter the F1 hybrid became the mainstream of cocoon production in Japan. Consequently, the improvement of silkworm had intended to establish the parental pure races for F1 hybrid.

As a result, sex-limited races by radiation breeding and oligophagous race that grew by the diet excluding mulberry leaf were established. Recently, the breeding goal has been shifted from the standardised raw-silk production to the extraordinary purpose. The idea is based on the practical use of biological function that is unrestricted by raw-silk production. Here, the races improved in such situation are introduced. But it must be notified that the races were excluded from the genetic stock.

Virus resistant silkworm race

The healthiness of silkworm is an important theme in the breeding. By the screening of the silkworm-genetic stock, the densonucleosis virus (DNV) resistant line was discovered. As a result of the genetic analysis, it is demonstrated that the resistant property is controlled by the single-dominant gene. Therefore, it is possible that the resistance expressed at the F1 hybrid. Consequently, the resistant race "Taisei" was successfully established.

Generally, thin and fineness silk thread is fitting for the making of the high-grade textile. And it was established that the race, Akebono, which formed 2.2 denier of silk thread. The thread was twisted together with chemical fiber, and the product which is called as Hybrid Silk has been sold. In 1998, Hakugin race was established as the race that had the silk fiber thinner than Akebono. Surprisingly, the thread size is 1.6 denier. It is the first record by the thinner size less than 2 denier in spite of tetramolter. It is expected that the super-thin textile made form the raw silk of this race attracts the attention of the worldwide as a particular product of Japan.

In contrast to the upper paragraph, there is a demand of the thickly and short raw silk for Western clothes, and an race which produces the silk thread of 4 deniers has been improved. A characteristic line, the original line of the race, had been established in 1952, though most researchers were not interested in the line. But the line was remained as a private collection until recently. The fact indicates the importance of genetic resources. Presently it was continuously preserved and registered as a collection of the genetic stock.

Colored cocoon and double cocoon

Colored-cocoon race has well known in worldwide. The yellow pigment of the cocoon is presently attracted because of the antibacterial activity of flavonoid that is a component of the pigment. Besides, the pigment of the colored cocoon is chemically fixed, and the utilisation of colored raw silk has been attempted. Hand spun silk fabric, which is called Tsumugi, formed from floss silk is very popular and traded expensively in Japan. The double cocoon, as an ingredient of the floss silk, is too scarce to use large amount, because the mounting technique of matured silkworms has been improved. So, double cocoon race has been established.

Transgenic silkworm

Tamura succeeded in generation of the transgenic silkworm in NIAS. Since exogenous protein gene could be integrated into genome by using of the technique, it will result in an increase in number of the genetic stock of silkworm.

Problem of the retirement from silkworm preservation

Presently due to the reduction of Japanese sericulture, it is difficult for the producer of silkworm egg to make an adequate profit. Some private companies retired from the egg production, and original races were lost before we knew it. Other companies are continuing the production by the moving of the location to foreign countries. In some prefectures, the section of the experimental station in relation to sericulture was abolished owing to the reduction of the farmers. In lucky case, the silkworm races possessed in such experimental stations has been donated to NIAS.

Koishimaru phenomenon

Some pieces of silk cloth were excavated from remains that seemed to be of ancient Japan around first or second century. When a copy of the ancient-silk cloths had been reproduced as an activity of archaeology, the utilisation of the silkworm race, which was same to ancient race, was aspired. Consequently, Koishimaru race, which has been known very well as the silkworm presently reared by the Empress, was chosen for the project. The news reports of the event made Koishimaru the most famous race in Japan. This race was established about 150 years ago and used by 1900s. Thereafter, the race was took the place of F1 hybrids.

The race has a little amount of silk and some defects on the silk filament except that the filament is slightly thinner than recent common race. However, it is owing to the brand image in relation to both the imperial silkworm and high class textile that the race has been the fashion in the nation. Since the people love the texture of ancient silk, further improvement of the race is hated. So the raw-silk producer requests farmers to produce the cocoons of the original race. As a conclusion, old races should be preserved for the demand in future

Silkworm Germplasm Database

The genetic-stock database of NIAS has been constructed and opened to the public by using of world wide web (WWW) on Internet. The primary characters which described above and photographs of egg, larva and cocoon were carried in the database. Unfortunately the data is written in Japanese language. Most of research reports which has been become the basic knowledge of Japanese researcher were also described in the Japanese. The problems related to the language, the conversion to electronic data and the linking up each data has caused to take time for the analysis of the data.

Management and Utilisation

The database was constructed by the data prepared in

1991. By now, some races add and some had been lost, however the database has not been updated. As the additional problem, the database is not a well-known fact for researchers in spite of searchable in the www search engine, for example Yahoo and so on. It is expected that further popularisation of Internet will increase the user of the database. As the encouragement of domestic sericulture was major purpose of the sericultural-research section of NIAS, the English presentation had not been active. Presently, NIAS is promoting to make English papers.

Characterisation of the Silkworm Genetic Stock

The genetic stock is categorised into four groups, native, improved, mutants and intermediate stock. Native race is a pure line, and the farmars are also able to produce the eggs by themselves. But many improved races are F1 hybrid which is functional only in the generation. The mutants were reared for the preservation of the mutant genes, not of gene pool. The breeding resource is fully improved race that will be registered as a new race in near future.

Native race and improved race are respectively divided into three types, Japanese, Chinese and European. The improvement is performed in order to make the races, Chinese and European, adapted to the climate of Japan. As the races were pure line for the production of hybrid, they were preserved in the condition to keep genetic distance against to Japanese in order to maximise the effect of heterosis. Typical features of the races are shown in Table 2.

Evaluation

In present Japan, any silkworm race could be reared without legal controls even if the race had problem. By

Table 2. Characteristics of geographic races of silkworm genetic resources conserved in NIAS

Geographic origin		*Japan*	*China*	*Europe*	*The tropics*
Voltinism		univoltine and bivoltine	univoltine and bivoltine	univoltine	polyvoltine
Larva	body marking	marked	plain	marked	plain
	body shape	slightly long	slightly short	big	thin and small
	growth	slightly slow	comparatively fast	slow	fast
	healthiness	slightly polyphagous slightly sensitive to pathogen	torelant to high temperature torelant to pathogen	sensitive to high temperature sensitive to pathogen	torelant to high temperature torelant to pathogen
Cocoon	shape	bale	ellipse	long ellipse	spindle
	color	white and extremely light brown	white and yellow	white and fresh color	yellow, green and white
	silk amount	slightly large	slightly little	large	very little
		filament size	slightly thick	thin	thick thin
		filament length	short	long	short very short
	others	double cocoon	good reelability	high degumming %	low raw silk%

1998, the controls existed in order to permit the distribution as a commercial race. The details of the requirement to register the race were described below.

— *Healthiness*: The race is able to be tolerant to the climate high-temperature and humidity in summer season. The mean value of the ratio of pupal number to larval number has to be higher than both control race and the mean value of 10 races those have been registered most recently.

— *Amount of Cocoon harvested*: The mean value of the amount of cocoon harvested have to be higher than both control race and the mean value of 10 races, that has registered most recently.

— *Cocoon shell weight*: It is desirable to be abundant, because the quantity percentage of raw silk is proportional.

— *Silk fiber size*: It is better, as it is thinner.

— *Silk thread length*: It is better, as it is longer.

— *Reelability*: It has to be higher than 73 percent.

— *Neatness defects*: It is better, as it is little. It has to be over 93 points on the defects.

— *Degumming loss of cocoon shell*: It is better, as it is little. It has to be lower than 26 percent.

— *Number of eggs laid*: It is desirable to be both abundant in the number of eggs and high ratio of the number to fresh cocoon weight.

At present, new race can be provided to sericulture, even if it is worse in any characters. However, the race patent has been acquired, since the system that protects the right of the breeder.

Sharing Information

The global sharing of the information about the genetic stock has been promoted in NIAS. Of course the translation of the data from Japanese to English is also in progress. It is necessary to unify the standard on quantitative character and technical terms, when the information of the genetic stock is exchanged.

Previously, Japan had provided the silkworm races for India and so on. The races were exchanged equivalently with same number of silkworm races. It seems hard to provide silkworm race one-sidedly. Since NIAS make Japanese sericulture to have competitive power, the unique races will not be exported to any other countries. That is depending on the sericultural trend in Japan.

Anybody considers obtaining silk fabric at low price even in Japan. And, the profit increases more for the silk reeling industry, when the raw material cocoon has gotten at low price, too. Though the quality is very high on the cocoon produced in Japan, it is very expensive. Therefore, it is advantageous of that the cocoon is imported from foreign countries, if it intends to produce high-quality silk fabric at low price. Sometimes the quality of the foreign cocoon has been bad to weave fine clothes Japanese traditional. So some private companies has chosen the method that they rear the Japanese silkworm in foreign countries and import the cocoon at low price.

If the genetic stock is reared in foreign country, the silkworms may not adapt to the foreign climate. Therefore, the developing country without the breeding technology may not be able to utilise effectively the genetic stock. In the meantime, the one-sided acceptance of silkworm race to Japan is seemed to be possible, but it is selfish. It is possible to share the maintenance of silkworm races owned by foreign country. When the necessity was occurred, then the country depositing the race could be allotted the eggs. If the races could be also used in Japan according to the contract with the country, it would benefit NIAS.

Recently, the accumulation of waste which does not decay in the natural environment becomes a problem. The raw silk is convenient fiber which never remains due to decay easily in the environment, from the standpoint. And, it is the only long filament got from the nature. Therefore, it is possible to consider that it is useful to keep the environmental. Accordingly, people may prefer such race rather than the race which produce highly heavy cocoon in spite of being often illness as some races in Japan. For the developing countries, the race which is healthy and tolerant to the change of environmental condition may be recommended. The stable production of cocoon is important than the pursuing of cocoon productivity.

Conservation

The domesticated silkworm, *Bombyx mori* has widely differentiated into a various races and strains under the long history of sericulture. After understanding of Mendel low, the discovery and conservation of mutants have also carried out and now more than 420 hereditary traits have been analysed and mapped on twenty seventh linkage groups. In Japan, many institutes (for example University laboratory, National sericaltural station, private company for silkworm seed and prefecture sericaltural station, so on) have been stocked those mutants. They were maintained through world war I and II.

Japan lost many things during world war II, but almost of silkworm strains were kept due to the efforts of the staff working at that time. Unfortunately, however, available institutes for the conservation of Silkworm Genetic Resources are decreasing in now due to the decline of sericultural industry of Japan.

Kyushu University and National institute of Agrobiological Science of Japan (NIAS) function as center for silkworm genetic resources in Japan. The genetical resources of NIAS include many kinds of mutants and pure lines derived from various regions and nations. On the other hand, Kyushu University completely focuses on the conservation for genetic mutants, and the number of those about genetical mutants is larger than those of NIAS. It seems that we can see more than 90 percentages of mutants now being in the world in about 460 strains maintain at Kyushu University

Besides of these two institutes, Hokkaido University, Shinshu University, Tokyo University of Agriculture and Technology, University of Tokyo, Kyoto Institute of Technology, Nagoya University, Institute of Sericulture and so on make an effort to maintain peculiar mutants, respectively.

Closed line culture system is employed in stock maintenance, because difficulties due to inbreeding depression are encountered. Usually, for maintenance of one line culture, three hundred larvae derived from 15-30 batched eggs are rearing in spring, when best season for rearing. In northern area of Kyushu island where Kyushu University locates, it is possible to rear silkworm from middle of April to middle of November. Thus long available period for rearing is enable to culture five generations in one year. This is advantage for genetical experiments. In main island of Japan, Honshu island, rearing season is shorter than Kyushu island.

The genetical traits (characteristics) of each strain are checking at all developmental stages by expert staffs, especially it is doing strictly at last instar larval stage and pupal stage to prevent genetical contamination. So far, method of long-term preservation more than one year has not been successful in this insect. For maintenance of silkworm strain, it is therefore necessary to rear each strain year after year. When users want to get some mutant for research or culture, it is necessary to order delivery of eggs before the start of spring rearing. Because, rearing for maintenance is usually done only one time in spring.

Development of Germplasm Resources

Recently two original methods for silkworm transformation have developed, and then transgenic lines by using those methods are accumulated rapidly. It is serious problems to maintain new transgenic line, because rearing is hard work. As the means of solving this matter, it is noteworthy the artificial insemination using a cryopreserved sperm established by Takemura et al.

Germplasm Database

Genetic informations of *Bombyx* mutant stocks maintained in Kyushu University had published as the notebook or the catalogue. Now, it is planning to reconstruct them as the databases with fine photograph (egg, larvae and cocoon and remarkable trait) display the traits in each strain. In NIAS, the database for genetic resources has been opened on Web.

Promotion of Global Sharing of Germplasm Information

Developing countries are calling for assistance or information in sericultural techniques including germplasm information. On the other hand, maintenance of germplasm for mutations in Japan and Korea is getting difficult due to the decline of sericultural industry. It is timely therefore that FAO and ISC organise this meeting. In order to realise the global sharing of germplasm information, it is necessary to establish working group to discuss rule on partnership for sharing.

Since the germplasm information already recorded at each institute are expected that those were written by various format, it will make difficult for users to understand real traits they want to search.

At present,

	Merits	*Demerits*
User		Hardly to get information for genetic resources at the same time.
Offer	Keeping high secrecy	Limitation of spread.
FAO		Difficulty of service.

If FAO plan will be available.

	Merits	*Demerits*
User	Enable to get information at the same time.	
Offer	Spread widely	Can not keep secrecy
FAO	Contribution to user	Increasing of tasks.

There are approximately 250 million animal species in the world and most of them are insect species. Among these insects, honeybee and silkworm have been related closely with human. Especially, silkworm is one of the representative agricultural insects and it has contributed greatly to the cultural development of human beings, particularly by providing clothigs. Thus, the value of silkworm resources is under national protection, as it is considered as one of agriculture related resources according to genetic resources management rule.

Sericulture in Korea, which began in 1900, has been developed and supplied many silkworm strains. Sericulture once greatly contributed to national economy as a major export product. However, it has been declined partially because population in the rural areas has been reduced. Now, sericulture is on the new age. Development of silkworm powder for blood glucose lowering agent and mass production technology of Dongchoonghacho are parts of this trend. Extension and conversion of traditional sericulture to a new era has been possible due to well-conserved silkworm genetic resources.

Non-cocoon forming strain for mass production of Dongchoonghacho and large body-sized strain for making silkworm powder has devoted to the development of such a new demand for sericultural industry. Moreover, development and conservation of silkworm resources is essential because they will be used as raw materials for current and future science, such as biotechnology. After the Biodiversity Agreement in 1993, each country has established it's own genetic resources conservation policy. So, collection and import of genetic resources become difficult, and safe conservation and systematic management of these resources have been necessary.

Conservation of Silkworm Genetic Resources in Korea

For the safe conservation and insurance of silkworm gene resources and in adherence to national protectionism according to Convention on Biological Diversity and utilisation for basic and practical study, the Dept. of Sericulture and Entomology now conserves 340 silkworm strains. These 340 strains (321 registered strains and 19 unregistered strains) were classified largely by their origin, voltinism, moltinism (Table 3).

Table 3. The present status of conservation of 321 silkworm strains

	European race	Japanese race	Chinese race	Tropical race	Korean race	Non classifed	Sum
Univoltine	32	-	-	-	-	-	32
Bivoltine	-	88	96	-	4	-	188
Multivoltine	-	2	-	9	-	-	11
Others	-	-	-	-	-	90	90
Total	32	90	96	9	4	90	321

The criteria for classifying silkworm are based on their origin, chemical characteristics. In RDA, the classification method based on silkworm's origin is used as the standard method when examine the characteristics of cocoon. Silkworm strains can be largely classified into five major strains: Japanese, Chinese, Tropical, European, and Korean.

8

Silkworm Germplasm Resources

Sericutlure in China can be traced back to 5000 years ago. In China, sericulture, basically the agro-based cash crop is practiced in over 18 provinces where the climate and rural backdrop vary considerably. The phased development of sericulture is oriented towards specific direction given in the successive ten 5 year plans of the country and made it as a sustainable enterprise. In 1984, the national Sericulture Research Institute had awarded one million yuan fund to carry out the research on Silkworm Gene Analysis, Establishment and Enrich of Silkworm Germplasm Resource, and Application of Mutation Genes as a key project by National Natural Science Foundation in the 7th 5-year plan.

Silkworm Germplasm Resources

Sericulture is supported by the silkworm gene resource diversity. Sericulture advancement are largely based on the silkworm variety improvement, and breeding of new silkworm variety in turn based on the material supply of rich silkworm germplasm resource. Owing to the long history of sericulture practice and wide diversity of geographical conditions, there is a very rich resource of silkworm germplasm bank in China. At present there are over 600 strains preserved in the central germplasm maintenance centre.

As the progress in silkworm germplasm collection and investigation, many of the strains are used in silkworm breeding for commercial activity and played significant role in the advance of commercial silkworm varieties in China. For Example, the pure lines Huahe, 672, 652, 303, Su17, Dongfei, Su12 were selected from the germplasm bank to be used as the parent materials for the breeding of three way hybrid "Su3.

Qiu3 × Su4" which had been distributed in Jiangsu and Sichuang province about 16million boxes from 1979 to 1988, and it was estimated by a net increase of 7million yuan for its better performance. Other dominant hybrids such as Yunhan × Hua9, Su16 × Su17, Dongfei × Huahe, Jingsong × Haoyue, 306 × Huashi, Dong34 × Su12, are all bred by the utilisation of purelines preserved in the germplasm bank, It was these hybrids that four times renovation was realised in silkworm varieties in China during the recent 40 years, and finally came to keep up with the Japanese level.

Since the discovery of heterosis in silkworm breeding, hybrids were used almost all over the world in commercial cocoon production, which had greatly enhanced both in quantity and quality of cocoon yield per unit. Meanwhile, there is the tendency of similarity in main economic characters and close in genetic relationship, which resulted in the elimination of vast local varieties with significant ecological diversity and the lost of some specific genes. Therefore, ecological adaptability of the improved varieties were greatly decreased, and the risk of epidemic of certain disease increased, that is the genetic-erosion.

One of the evidence is the decay of sericulture in Europe in 19th, that epidemic of the pebrine disease had ruined European sericulture due to the pebrine-sensitivity of silkworm varieties used during that period. Sustainable development of sericulture industry is largely depending on breeding new silkworm varieties with one or more distinctive character by the effective utilisation of silkworm germplasm resources, those are the diversity of germplasm materials could always supplement the new genetic resource for the breeders.

The utilisation of silkworm gene mutation in commercial production is widely noticed by silkworm breeders. First of all, some of the mutations in economical characters can be directly utilised in silkworm breeding. The polyphagous mutation had been utilised to breed silkworm varieties adaptable to artificial diet; the trimolter mutation was utilised to breed trimolt silkworm varieties to produce especially fine silk; the non-glutinous mutation was used to produce natural loose eggs. Secondly, some of the morphological characters have multiple functions that also positively effect on the economical characters, therefore, it is possible to be directly utilised in new variety breeding.

Experiment had showed that *K* gene had positive effect on feed efficiency and heterosis; *PS* and black pupa (*bp*) also showed the potential to increase feed efficiency; the introduction of non-scale mutation on the wing (*nlw*) can greatly improve the silkworm seed production condition. Thirdly, Sex-control can raise single sex larvae, such as to rear only male silkworm in commercial cocoon production.

There are many practical methods developed to control the sex balance of silkworm under the joint effort of scientists from different countries, these are:

androgenesis, parthenogenesis, balanced—lethal mutation, and control of incubation condition. It was discovered by Chinese scientists that the sex-linked chocolate gene mutation (*sch*) was recessive gene that is very sensitive to high temperature during incubation stage. When genotype ZW crossed with $Z^{sch} Z^{sch}$, the female embryos of their hybrid could be killed during the incubation stage, therefore, only male silkworm will be hatched.

Many genetic discoveries in silkworm are belong to the forefront of genetic research, and enriched the genetic theory development. It was estimated that silkworm genome consist of about 4.8×10^8 bp, its genetic information volume is about one sixth of human being. There are over 450 morphological, physiological and biochemical characters recorded at present, among them 300 (including multi-allele) had been located on 27 groups of the total 28 chromosomes. This is one of the most detailed linkage map in the agricultural biological resource.

Silkworm is an excellent material as the experimental animal. It is well known fact that advantages of silkworm as genetic experimental materials. For example: the chocolate colour mutation of newly hatched larvae, there are 6 mutations had been discovered in this character, that is common chocolate colour (*ch:*13-9.6), light chocolate colour (chP: 13-9.6), second recessive chocolate colour (ch-2;18-0.0), dominant chocolate colour (la: 9-22.1), sex-linked chocolate colour (sch:1-21.5), maternal chocolate colour (cm: 20-21.9), plus the artificial created system of sex-limited chocolate colour (W.+sch/Zsch). Among these mutations, there are dominant and recessive inheritance, sex-linked and sex-limited inheritance, and maternal inheritance, in addition, some of them are distributed in same chromosome, some in different chromosomes (belonged to 4 linkage groups). Therefore, it is an excellent material for genetic experiment for student genetics courses.

Collection, Assortment and Identification

Investigation on germplasm varieties is one of the most important part in the maintenance of silkworm germplasm resource, only based on these basic research results that the germplasm could be effectively utilised in silkworm breeding and genetic research. Collection, assortment and identification are usually done in the early stage, and maintenance was strictly followed the specific rules in China, that is:

— Those newly collected varieties should be separately raised during the first two generations when newly introduced, and could be raised in the same room of the existed stocks only after confirmed that they are pebrine disease free.

— The uniqueness of each variety is maintained only through systematic and scientific manipulation of the germplasm. Otherwise, it will degenerate to loose its original characteristics in genetic markers or in its performance of economic characters, or it will become lost because of drastic reduction or fertility to the extent that it is no longer able to produce viable offspring.

— Generally, rearing once a year in the congenial season for monovoltine and bvoltine races, two or three times for diapaused multivoltine, nonstop rearing for non-diapaused multivoltine.

— In order to prevent inbreeding depression and maintain the specific properties, the newly hatched larvae should be collected from at least 20 layings:

cut one corner of eggs from each of the 20 layings, and stitch them together, brush 1500-2000 newly hatched larvae in the form of mixed rearing.

— Strictly maintain the purity of the variety in each stage of egg incubation, larvae rearing, mounting, cocoon collection, moth emergence, and egg production. In any case, if the abnormal character which is different from the original record, delete it, or separate if it could be a new mutation.

— Duplicate set of eggs should be produced so as to prevent the distinction of the stock in case of unsuccessful reproduction due to some unexpected circumstance.

Investigations on the silkworm germplasm resources are mainly including following aspects:

Records of biological features and characteristics

— Race, voltinism, moltinism;

— Eggs: shape, colour, shell colour, coherence;

— Larvae: taxis, eating habit, marker, body shape, colour; size;

— Cocoon: size, shape, colour, wrinkle;

— Moth: size, colour, mating habit, egg laying habits, laid egg number;

— Except the word records, photocopies of egg, larvae, cocoon, pupa moth should be maintained.

Investigation of economic characteristics

— Characteristics investigated during rearing: development uniformity, duration, survival rate of larvae, larva-pupa survival rate, rate of dead worm cocoons, cocoon weight, cocoon shell weight, cocoon shell ratio.

— Characteristics should be investigated by special inspection machine: filament weight, filament length and size, reelability, neatness, strength and elongation, cohesion, degumming loss percentage of cocoon shell, lousiness and etc..

Identification of resistant ability

Resistant ability is generally investigated under the artificially set environment condition to identify the specific property in different adverse conditions. For example:

— To identify the tolerant ability to high temperature and humidity, larvae reared under the environment at the temperature 30-32°C, and RH over 90%;

— To identify the resistant ability to certain disease, inoculation of a specific bacteria or virus.

— To identify the tolerant ability to certain insecticide or fluoride pollution, by oral administration of miro-concentration of insecticide or fluoride.

Genetic analysis

— Gene behaviour and gene location of the new mutation. There are high rate spontaneous mutation in silkworm, to identify and study its function, inheritance and location, enrich and modify the silkworm gene linkage map, are one of the important aspect of germplasm maintenance.

— Biochemical genetic research. To study isoenzyme pattern of hemolymph, skin, midgut and silk gland, protein type of blood and digestion juice in different development stage are also widely concerned in germpalsm investigation.

— Molecular linkage mapping by bio-technique such as RAPD, RELP, SADF, AFLP. This is the newly developed field and also widely recognised in silkworm germplasm research, will be the hot spot in the near future.

Long Term Cold Storage Techniques

The polyvoltine varieties have to be raised for many times a year in the germplasm maintenance, which not only created more works for the maintenance but also resulted in fast degeneration for more rearing generation in a year. The new cold storage technique had made it possible to rear diapaused polyvoltine varieties once a year. That is to rear the polyvoltine varieties in summer season, laid eggs (August 1-5) are preserved under 25°C for 35-40 days, then under 20°C during September to October, after that, the temperature gradually decreased to the natural temperature until 5°C, egg bath takes place in mid-December, and then be transferred into cold storage under 2.5°C or 0°C, take out for intermediate temperature under 10°C-15°C for 7 days in late April, and put back to the cold storage under 2.5 °C until mid June before incubation starts.

Silkworm will be reared again in summer season. Experiments had shown that there were no significance difference between the new method and normal rearing cycle in the rearing performance such as hatching rate, cocoon weight, cocoon shell weight, shell ratio, survival rate of larva-pupa, and voltinism character.

Maintenance of lethal genes

There are many lethal genes discovered in silkworm, and

these genes require special techniques to be maintained in the germplasm resource. The small egg mutation *sm*, when the female moth is an homozygous *sm*, it lays 100% small eggs which are only half of the normal size and normal embryo cannot be developed and become dead eggs. Usually, to keep this gene, rear 5-10 layings of heterozygous *sm* separately, and select the layings with a ratio small egg: normal egg 1:1 as the seed layings, and 5-10 laying are reared separately again at the next generation.

This will cost much more man power and material for maintenance, and moreover, very hard to be used for linkage analysis as marker gene. Now, take the dominant gene of larval zebra marking, which is in the same linkage as the marker, make the Ze +/+sm♀ × + sm/ +sm♂ system to keep the sm gene, just use zebra marking × non-zebra marking can produce the heterozygous *sm* eggs, therefore, only one batch of mixed rearing is required for the next generation. Here, the complete linkage of female silkworm was used, it is easy to distinguish that the zebra marking individual is heterozygous *sm*, to produce normal eggs, and non-zebra marking individual is homozygous *sm*, to produce small eggs. This principal can also be used in the maintenance of second small egg *sm-2*, kidney shape egg *ki*, shriveled egg Gr^{c01}, by the system of ++/*ch sm*-2♀ × *ch sm-2*/ *ch sm*-2♂, E +/+ki♀ × + ki/+ ki♂, + +/pGr^{c01}♀ × pGr^{c01}/pGr^{c01}♂. Other lethal genes like larval lethal gene *l-nl*, small non-molting larva *nm-d*, white dead egg *al*, stinky larva *sku* etc. can also make similar system to be maintained in a much easier method.

Present Status of Germplasm Maintenance

Since 1949, there is a great achievement not only in the collection, assortment and investigation of local varieties all over the Chına, but also introduced a lot of new silkworm varieties from Japan, former Soviet Union, and other countries to enrich the silkworm gene bank. There are over 1000 strains or varieties of mulberry silkworm preserved in different institute in China at present, which including five aspects of the sources:

— Local varieties;

— Improved varieties;

— Introduced varieties;

— New strains of gene mutation;

— Wild species with close genetic relation.

All these preserved resources could be divided into five different ecological types, that is:

— Chinese temperate race;

— Chinese subtropical race;

— Japanese race;

— European race;

— Tropical race

There are 1000 varieties or strains (including mutations) preserved in Sericulture Research Institute of China Agricultural Science Academy, in Zhengjiang City of China. 300 breeding materials and 602 gene mutations are preserved in College of Sericulture, South-West Agricultural University, Chongqing City of China. And some other breeder's stocks maintained in different universities and provincial level institutes.

Achievement

Sericulture Research Institute of China Agricultural Science Academy was awarded for Second Prize of Scientific Advancement on Achievement of the research and application of silkworm germplasm resources in 1995. Their main achievements included following aspects:

Finished the gene survey of silkworm germplasm in China: Systematical investigation on the gene bank of silkworm germplasm varieties was completed while keeping the target to discover and collect different types of gene mutations. In the national silkworm germplasm centre (Zhengjiang), over 20 character had been investigated including morphological, physiological and biochemical characters of each variety such egg colour, egg shape, chorion colour, newly hatched larval colour, body shape, cocoon colour, cocoon shape, pupa colour, pupa shape, moth colour, moth shape, moth eye characters, globular cell of hemolymph, amylase of digestion juice, amylase of blood, albumin, of hemolymph, low molecular lipoprotein of pupa, eating habit of the whole preserved varieties (around 1000 accessions) in its germplasm centre during 1985-1995.

Meanwhile, 400 mutation individuals were discovered, among them 126 gene mutations were selected out and 50 were discovered as new mutation in the world. During the corresponding period, there was a total around 90 new mutations on mulberry silkworm discovered in the world, therefore, this institute alone had account for more than 50% of the total new discoveries. All these research discoveries had greatly enriched the silkworm germpalsm bank in China and supplied the materials for the genetic research and silkworm breeding.

83 gene mutations were analysed: These 83 gene mutations were determined in loci, and its linkage group. Among them, 26 gene mutations are discovered and localised at first time in the world, such as Zhengjiang gray egg (Gr^{Z}, 2-6.9), Xu gray egg (Gr^{Xu}, 2-6.9), second recessive chocolate colour of newly hatched larvae (*ch-2*, 18.00), Zhengjiang narrow thorax (*nb-Z*, 19-12.0), Hainan flossy cocoon (*Syc*, 7-32.1), Jiang's striped marking (P^{JS}, 2.00), black larva (*blb*, 9-22.9), second dark colour of moth (mln^2, 18-14.1), second face and tail marker (cts^2, 16-4.6), Sheng's black marker (P^{BS}, 2-0.0), oe translucent larva (*oe*, 10-?), densonucleosis-resistant gene (to Zhengjiang strain) (*nsdZ*, 15-?), non-lipis wing of moth (*nlw*, 13-?), second recessive gray egg (gr^c, 12-?), 013 translucent larvae (*ox-4*, 5-?).

Determined the marker genes or representative genes for the total 28 linkage groups of silkworm Bombyx mori and modified the 18 linkage: Marker genes for the 28 linkage groups were determined, its corresponding linkage retrieve system of the 28 linkage and detecting system for gene loci were established. The map of the 18th linkage group was modified based on the research result on genetic analysis of new gene mutation, the second recessive chocolate colour of newly hatched larvae *ch-2*, by the application of the above established system. This modification was acknowledged and adopted quickly by the academic circles of the international sericulturists.

Some valuable mutations were introduced into the improved varieties to breed practical varieties; a number of breeding materials and basic lines with specific characteristics were obtained through utilising the mutation genes:

i) Normal colour of silkworm moth is brown-rice colour, it is very difficult to distinguish the Japanese or Chinese race for mating during the hybrid production. Research had approved that the new mutation mln^2 had no negative effect on the economic characters, therefore, it was introduced into the improved variety "Fengyi" (Chinese race) to make the basic variety "Fenghei" with dark moth character. Its moth is deep dark colour, and very easily be recognised from normal colour moth, therefore, when mated with normal Japanese variety, it is so distinguish in colour that any pure pair could be caught by eyes right away, and 100% hybridisation can be completely ensured.

ii) The non-lepis wing mutation (*nlw*) was discovered in 1987, and localised in 13 linkage group. By introducing this character into improved variety, moth dust could be decreased by 95% and the production environmental condition could be greatly improved.

iii) Disease caused by parasitic maggot has been the one of major lost in cocoon production in the tropical areas. It was found that maggot fly seldom laid eggs on the black body larva with the black body gene mutation (*blb*). This gene mutation has also introduced into improved variety, thus could supply basic material for this particular breeding purpose.

By utilising the morphological characters to label the Densonucleosis-resistance gene *(nsd-Z)*, a new method for breeding Densonucleosis-resistant silkworm varieties has developed. Linkage analysis had shown that the gene were closely linked with *nsd-Z*, desonucleosis-resistant varieties D01 (Chinese race) and D02 (Japanese race) were bred. By utilisation of D01 and D02, it is easily to select DNV resistant individuals by the marker gene bl2 without DNV administration.

Polyphagous gene mutation was found both in Chinese race and Japanese race, and four strains with polyphagous gene were bred in Chinese race and Japanese race respectively. It is the first time to found the polyphagous gene in Chinese race in the world. A gene bank that is the most complete one in China was established and corresponding database by FoxPro 2.5 completed. Some specific preserving techniques for lethal genes were developed. In order to optimise the structure of gene bank and decrease the risk of gene disappear, average 6 marker genes were carried by each preserved gene strain, and each gene mutation was carried by at two strains.

South-West University has been mainly focused on the research of egg colour mutation, radiation induced gene mutation and molecular linkage map. Since 80th, over 10 new gene mutations were discovered, and 220 marker genes were preserved, some valuable gene mutations were found, a total 602 gene mutations were maintained in this University. Dai F.-Y. et al. reported that A new mottled translucent *oh*, was discovered in the preserved system of silkworm in the Key Laboratory of Sericulture of the Ministry of Agriculture, it resulted from a recessive mutation in the 20th linage group, its gene locus is at 20-0.0 and *ghe* gene appears to be dominant over *oh*, this mutant is now tentatively designated with the gene symbol oh^{m} and was found to have an epistatic effect over Chinese translucent (*oc*).

Lu C. et al. found a citron-egg gene (*ci*) that belongs to the 20th linkage group for it linked with an hoarfrost translucent (*oh* 20-0.0) gene which is a marker gene of the 20th linkage group. *Ci* gene is located at the locus 20-21.9. Lu C. et al. (1991) found a spontaneous mutant of the new ellipsoid egg from *Ng* system. Its morphological appearance is similar to that of *elp*, but it is longer and thinner. Genetic analysis indicates new ellipsoid *eff* is controlled by a recessive gene that is independent of gene *elp*, named as *elp-2*.

Another spontaneous mutant or the chocolate purple has been discovered from the white-5 eggs system in 1990. Its morphological characters and hereditary mode are different from *Ia, sch, cm, ch, and ch-2.* The gene is marked with ch^p which is the allele of $+^{ch}$ and *ch*. Xiang Z.-H., studied the relationship between cocoon fluorescent colour (*FC*) and varieties, sex, mounting environment and the inheritance pattern of *FC*, found that the pure yellow type was identified to be incompletely dominant to the pure purple type, cocoon fluorescent colour is controlled by multiple genes, there are at least two pairs of genes (*Yf-1* and *Yf-2*) that controlled the yellow colour and at least two pairs of others (*Vf-1* and *Vf2*) that controlled the purple colour, the homozygous *FC* cocoons were less affected by environmental temperature and humidity, but the heterozygous *FC* cocoons were remarkably affected.

As above-mentioned, there is a total 26 units involved with silkworm germplasm resource maintenance or research. Except the above two major institutes, the relatively important ones are: Sericulture Research Institute of Zhejiang, Guangdong, Guangxi, Hubei, Hunan, Shuaanxi, Shangdong, Hebei, Xinjiang, Liaoning; Zhejaing University, Suzhou University, Southern Agricultural University, Huasheng Group of Sichuang Province, Qingzhou Egg Station, and some parent egg stations. There are over 100 silkworm varieties mainly in local varieties, special varieties and breeder's stock maintained in these institutes.

Fox example, Sericulture Research Institute of Zhejaing Province had introduced the balanced lethal system from former Russia, and characteristic with the breeding of only-male rearing varieties and parthenogenesis. Department of Sericulture, Zhejiang University had maintained some special varieties such as flat-way spinning, nudity pupae, extra-fine or superthick filament size, varied cocoon colour etc. Southern Agricultural University had maintained the tropical strains.

After a long period of continuous breeding research, several basal varieties with special characteristic were made. 651 and 652 are the basal varieties with especially long filament length; EH_{1new} with especially high cocoon shell weight, average 0.7g and shell ratio around 27%. "123" with especially high cocoon shell ratio, TB with extremely high in cocoon shell ratio (Its average cocoon shell ratio by batch reached 30.58%, the highest of female individual reached 31.79%, and male 38.09%.); "125" with high raw silk rate (around 28 %); "SG" with dominant trimolter (especially fine filament size, 2.0d).

Moreover, the basal variety breeding of rich egg laying and superthick filament size was one of the national silkworm breeding program in the eighth five-year plan. The task of this program is to breed a variety which could produce more than 800 eggs per moth or filament size thicker than 4.0d and cocoon shell ratio exceeds 22%. Chen K.-P. et al. had reported that "He" with hyperlaying was discovered from the 400 preserved silkworm variety resources in 1986, and it was used as one the parent material to breed the basal variety HEK which lays 821 eggs per moth average, the most 983, cocoon shell ratio reached 22.08%, cocoon weight 2.208, shell weight 0.486g, larval duration 23-25d. Wu Y.-C., et al. also had reported that four rich egg laying varieties—GE, GCL, GLZ, GLBZ had been bred from 1991-1995 by classical breeding techniques. The highest one, GE reached 1017 eggs per moth on average, and the highest moth 1083 eggs. The new varieties with filament size 4.5d and shell ratio 25.45 also bred.

Lin C.Q et al. investigated on 251 preserved silkworm varieties for their cocoon shell production efficiency, the average (cocoon shell/food ingested) was found to be 10.11%, highest 12.19%, lowest 6.71%, bivoltine was higher, and followed by monovoltine, then polyvoltine. Among bivoltine, Chinese is the highest, followed by Japanese, then polyvoltine. It was positively correlated with cocoon shell rate, cocoon shell weight, average cocoon shell weight per day in 5th instar, no correlation with the 5th instar duration of larval duration, rate of dead worm cocoons, cocoon weight and digestibility.

Moreover, 200 silkworm varieties were investigated on their resistance to densonucleosis through oral administration of the pathogen. It was discovered that 28 of them had the resistance. Two resistant lines were selected and the densonucleosis resistant gene *nsd-Z* and eye spot gene *bl* liked lines were created, so that it is possible to breed densonucleosis resistant varieties with eye-spot as genetic marker instead of administration of the pathogen to silkworms. Chen K.P., et al. found significant difference among preserved silkworm varieties to NPV-sensitivity, the resistance is of incomplete dominance to susceptibility, it is controlled by over two pairs of genes.

Lin C.Q. et al. tested the cocoon filament size of 200 preserved silkworm varieties in spring rearing season of 1992. The average size of all varieties tested is 2.80 dtex., the thickest is 3.93, and finesse is 1.72. The size of different strains is in the sequence of Japanese>European>Chinese>tropical in geographical races, and monovoltine >bivoltine> multivoltine in voltinism classification. Filament size is negatively correlated with filament length, percentage of sericin in cocoon shell and neatness, but positively correlated with cocoon shell weight, pupa weight and filament weight.

Lin C.Q. et al. investigated 209 preserved silkworm varieties on their cocoon filament properties, results showed that the average filament length was 840m, longest 1273m, shortest 344m, average reelability is 79.4%, highest 100.0%, lowest 41.5%, average neatness was 93.9, highest 99.2, lowest 76.7. It was also reported that resistance to fluoride and NPV are significantly different among the preserved varieties, the resistance to fluoride in the varieties of Fengyi, 872, 54A, Fenyi × 54A, Huafeng × Xuesong, 871 × 872 are stronger than other varieties, the resistance to NPV in the varieties of 318, 416, Qiu3, 317 × 318, 415 × 416 and 871 × 872 are higher than the other tested varieties, there is no relationship between the resistance to fluoride and NPV, but it is possible to combine both of the character in the same hybrid such as 871 × 872 and 317 × 318 which showed high resistant ability to both fluoride pollution and NPV.

Furthermore, deguming loss percentage of cocoon shell in boiling water for 30 min was tested in 250 silkworm varieties preserved, the highest and the lowest are 27.2% and 13.1% respectively, with an average 20.2%. As for voltinism concerned, monvoltine races is comparatively low and that of polyvoltine races is higher. It was positively correlated with sericin ratio of cocoon shell and cocoon shell ratio, but negative correlation with reelability. The average soluble sericin percentage of cocoon shell in tested races is 78.7% and the highest is 97.7%, lowest 50.3%.

There is no significant difference between female and male cocoons. Significant difference was found among different varieties to cold storage duration. No disadvantageous influence was found the two sex-linked genes of the balanced lethal line S-14 introduced form Russia on the viability and cocoon weight in the descendent generation.

Zhang Z.F., et al. reported that a new natural mutant: normal temperature lethal gene (*ntl*) was found from Chinese bivoltine native variety Yuhang 7. When the fifth instar larvae were reared the viability was 10.7% at the temperature of 23.3°C, and 90% at 29.4°C. The gene *ntl* is recessive, located at the 5th chromosome and very close to *oc*. Zhao Q.L. et al. discovered a mutant of translucent silkworm with medium transparency in the breeding variety Dong 3219, it causes a lethality of 21.16% during the first molting. This translucent character is controlled by a pair of recessive genes located on the fifth linkage group.

Except the national sericulture research institute, there are over 20 provincial institutes and university dealt with silkworm breeding in China. Chinese breeders had particular experience and techniques on the breeding of silkworm varieties adaptable to adverse rearing condition. A complete collection of thesis on silkworm breeding records, techniques and experiences will be published on October of 2002, in which 138 papers were included after 1949 in China. Besides, some silkworm varieties with special economic characters or for special purposes were bred. For example:

— Fluoride pollution has been seriously disturbed commercial cocoon production due to the rapid development of rural industry since early 80s of 20th century in China. In order to stabilise cocoon production in the major sericultural region, some highly fluoride-tolerant varieties were bred. $Huafen_{Gw}$ × Hue.A can normally developed when fluoride content reached 60mg.kg in the mulberry leaves during 1st-3rd instar and 120mg.kg during 4th-5th instar. Lu.Pin × Qing. Guang can grow very well when fluoride content was 100mg.kg during the whole larval stage. Other popularised varieties such as Qiufeng × Baiyu, Fengyi × 54A, Xia7 × Xia6, Xinhang × Keming and etc. are all relatively less sensitive to fluoride pollution.

— *Practical varieties with super-thick filament size:* The newly bred hybrid "Xinmiao × Mingri" can produce filament size 5.245dtex, and other characters are reach the popularised varieties, that is, cocoon weight 2.67g, shell weight 0.680g, shell ratio 25.45%, filament length 1094m, reelability 79.53% and non-broken filament length 870m. Its raw silk and the made fabric tests had shown a significant increase in bending rigidity and elasticity, tensile strength, elongation and elasticity, which indicated that its raw silk should be a suitable material for the exploitation of silk fabrics with better anti-wrinkle and hand-feeling.

— *Sex-limited cocoon colour:* Female cocoon is yellow and male white. It is easily for sex-discrimination and the male cocoon can separately be reeled to produce high grade silk. CY_{2c} (Chinese race) and JY_{628} is one of this type of commercial varieties. Their pureline performances are 1.543 and 1.694g in cocoon weight, 0.388 and 0.419g in shell weight, 25.15 and 24.71% in shell ratio, 83.09 and 93.05% in survival rate of larva-pupa respectively. The performance of corresponding hybrid was close to the control variety Chunlei × Zhenzhu, and its male cocoon produced much better quality silk.

The International Silkworm Molecular Breeding Plan initiated in the beginning of 90s, and Silkworm Genome Plan started at present by cooperation with Japan and other foreign countries. During recent ten years, progress on the molecular linkage study had been rapidly advanced, especially in the construction of molecular linkage, isolation of important molecular markers, gene location of important economic characteristics. The major part of the silk gene and silkworm mitochondria genes had been successfully sequenced, and gene clone, QTLs and transgenic research are all on the way.

The construction of molecular linkage are mainly depending on the molecular marking techniques such as restriction fragment length polymorphism (RFLP), random amplified polymorphism DNA (RAPD), selective amplification DNA fragments (SADF) and amplified fragments length polymorphism (AFLP) had determined 7 RAPD marker were linked with the morphological gene Y, Nl^1, Nd, L^c, L^{ca}, *od,* and *sch.* He N.-J., et al. had reported that a molecular linkage map of silkworm was constructed successfully based on SADF and RAPD markers. Xiang Z. H. et al. It was expected that a high dense molecular linkage map could be establish soon after combining with morphological characters and isoenzyme pattern.

Future Development for Silkworm Germplasm Resource

— to stable filial generation, maintain or store the mutation strains;

— to explain the genetic mechanism of gene information and expression;

— to study the new method and techniques for maintenance, management and multiplication;

— to exchange and manage the germplasm information system;

— to study the maintenance techniques by tissue and cells;

— to keep certain DNA gene fragment;

— to study the gene (or DNA) structure and function;

— and to exploit new materials.

Maintenance of silkworm germplasm resources is quite different from some other agricultural gene bank, that is non-Mendel population, but artificially maintained inside the door by plan. These silkworm germplasm resources are kept in a few institutes in several countries, and each institute has its their own special gene bank. In case of failure in one institute, it will disappear in the world. Therefore, an international cooperation net for joint maintenance of silkworm germplasm resources might be established.

Germplasm Database

The computerised information retrieval system for germplasm resource management could efficiently and accurately supply the complete data file of different stocks or target varieties for the breeders. This could be established only when relatively large germplasm resources had been maintained and been well investigated. The earliest Silkworm information retrieve system and silkworm variety database in Chinese output was establish in 1984 in China. In this system, there are 437 varieties included, which divided into five races: Chinese, Japanese, European, tropical and native breeds, and other breeds (sex-limited, trimolters, mutations).

There are 94 items could be searched for, such as: breed No., race, origin, institute name for maintenance, larval number per gram newly hatched silkworm, hatchability, larva-pupa survival rate, cocoon weight, cocoon shell weight, cocoon shell ratio, 5th instar duration, life duration, filament weight, filament length, reelability, filament size, neatness, laid egg number per moth, moltinism, voltinism, body shape, colour, marker, cocoon shape and colour, egg colour and egg shell colour, etc. The main functions of this data retrieve system are:

— Search for the maximum or minimum value of any variety or character.

— Search for the varieties which meet the requirements in all characters set by the user.

— Arrange any economic character in sequence according to the user's request.

— Search for any variety or any characteristics as request.

— Data output could be fully or partially printed in Chinese or code.

In 1995, a new silkworm germplasm database was set up by FoxPro 2.5 in the national Sericulture Research Institute. It can retrieve the information of mutation genes, gene strain, gene analysis, gene conservation, all characteristic of 1000 preserved silkworm accessions, and research data processing. It can be revised or increased if there is any new discovery. Unfortunately, there are all in Chinese. And the related information had not been uploading into Internet yet. Another data source for germplasm resource in China is the book, Record of Silkworm Varieties in China, edited by Sericulture Research Institute of China Agricultural Science Academy, 1987, Agricultural Publication Press, recorded 510 silkworm varieties as mentioned in 4.1. However, it is also in Chinese.

This is the major problem for foreign scientists to approach information on sericulture research in China though the tremendous achievement had been made. This situation might be changed in the near future, for more and more scientists on sericulture field came to know English very well. The new problem is lacking of fund to publish these information in English for the government does not encourage in this type of work.

Sharing Information in Germplasm Resources

It is very important to sharing information in silkworm germplasm resource, however, the most complete and well-studied germplasm is only maintained in a few countries, such as Japan, China. There are frequent information and visiting scholar exchanges between Japan and China, but not too much among other countries.

- The most difficulty part for global sharing of silkworm germplasm information is the language, especially in China. As already noticed, China had made great advance in silkworm germplasm research, unfortunately, most of the paper and books published is in Chinese, and similarly happened in Japan, most of the information available are in Japanese, both of the languages are not the common languages used in the international exchange.
- Difficulty to make connection. There is no existed pamphlet with the list of name silkworm breeders, fax, email etc.
- Lacking of experienced expert on germplasm maintenance. Generally the countries newly developed in sericulture, personnel dealt with the germpalsm maintenance are lacking of special knowledge and experience, which make hard to properly maintain the germpalsm, nerveless to say about exploitation of new resource.

Due to the unbalanced development of sericulture in countries, it is not possible to directly share germplasm resources. It is also not fair for unconditional sharing of some important materials while the owner country had spent so much human and material resources to developed it. However, international exchange in silkworm resources had been going all the time by one way or another. For example, China had use the hybrid rice resource to exchange a lot of silkworm resources with Japan during 80[s], and bought the balanced-lethal gene strain from Russian 5 years ago. Some of the genetic materials can be exchanged based on a common interest if it is within the permission of the country administration. The problem is

— What is really needed? Before having a clear knowledge about the germplasm resource preserved in certain country and considering practical conditions of the introducing country, it is not wise to take any silkworm variety.

— How to maintain the introduced resources properly. It had been always happened that the introduced new material were totally lost its original characters after a few year in some of the developing country due to lacking of knowledge and techniques for germpalsm maintenance.

9

Management of Silkworm Rearing

Sericulture is the technique of silk production. It is an agro-industry playing an eminent role in the rural economy of India. Silk fibre is a protein produced from the silk glands of silkworms. Five varieties of silkworms are reared in India for producing this natural fibre. *Bombyx mori*, the silkworm, feeds on the leaves of *Morus*, to produce the best quality of fibre among the different varieties of silk produced in the country.

Antherea assama produces the famous muga silk. This variety is confined only to the Brahmaputra valley of India in the world. Tasar silk is a product of *Antherea mylitta*, which feeds on *Terminalia tomentosa* grown in the thick jungles of Bihar, Madhya Pradesh and Orissa. *Antherea royeli* and *Antherea perniyi* has been recently introduced to produce the oak tasar silk, *Phylosamia ricini*, the eri silkworm, which feeds on *Ricinus communis*, is raised in Assam and Orissa.

Mulberry

Bombyx mori is reared on the mulberry, considering the economical conditions, such as rainfall and the nature of soil, different systems of plantations for raising mulberry are practiced in India. In the rainfed areas, mulberry is planted at a distance of 7.6 X 7.6 cm and under irrigation the spacing of 5 cm X 5 cm and the row system are followed. In India where the temperature ranges from 16°C to 31°C. The silkworm *Bombyx mori* are reared throughout the year.

The total life span of this silkworm is 50 days, out of these egg stage is 10 days, larval stage is 25-30 days and the pupa stage is 10 days. It passes 4 moults during growth. At the end of the larval duration, the silkworm emits silk from its mouth and constructs a cocoon on a scaffolding. The average annual yield of cocoons in India is as low as 150 kg under rainfed conditions, and under irrigated conditions it is about 400 kg. The cocoons are cooked in hot water and the silk fibre is unwound from the cocoons. This process is called reeling.

Ericulture

This silk produced by *Philosamia ricini* is called eri silk. It is mostly grown in Assam in eastern parts of India. The food plants of this silkworm is castor (*Ricinus communis*) The alternative food plants are *Heteropanax fragans, Manihot utilissima, Earica papaya, Ailanthus* sp., *Plumeria acutifolia*. This silkworm is miltivoltine and is reared indoors. The eggs are white, hatch in ten days. The hatched larvae are mounted on castor leaves in the rearing house and are allowed to grow by periodical feeding.

The worms moult four times during its larval period of 30-32 days. It is generally hardy and not susceptible to diseases. At the end of the larval period, the worms spins the cocoons. The cocoons are usually white. The cocoons of the eri silkworm cannot be reeled, as they are made up of uneven fibers. Generally, after the emergency of the moths, the cocoons are used for producing spun yarn.

Tasar Silk

For the extraction of tasar silk three species of *Antherea* are used in India. They are *Antherea mylitta, A. perniyi* and *A. royeli.* This silkworm is reared on trees of *Terminalia tomentosa, Terminalia arjuna.* Tasar silkworms are reared wild in nature. They are usually green in colour. The larvae pass four moults before they complete their duration. At the end of the larval period, the larvae spin a ring like structure around the twin and a along peduncle before it constructs the cocoons. The cocoons are large and brown or yellow. The cocoons are cooked in caustic potash and reeled to extract the fibre. Their are also spun for manufacturing coarse fibre.

Muga

Antherea assama produce golden yellow silk. This process is found only in the Brahmaputra Valley of India. It is semidomescticated. The worms, which crawl down at the end of their larval period, are collected by the rearers. They are allowed to spin cocoons in the rears' houses. The worms are raised on *Machilus bombycina* and *Litsaea polyantha* trees.

At the end of the larval period, when the worms are ready to spin the cocoons, they crawl down the tree in search of suitable place for the construction of cocoons. Such worms are collected and are allowed to spin the cocoons in 'jali' made of dried twigs. The cocoon has a very weak peduncle and is golden yellow. The cocoons are boiled in soap and soda solution and are reeled.

Mori-Culture

Today mulberry is cultivated in about 2.90 lakh hectares in India and the stress is upon productivity. Earlier the mulberry production per hectare was around an average of 12,000 to 15,000 kgs of leaf under irrigated conditions. Presently the average production has been tremendously increased due to the adoption of new technology like proper maintenance of mulberry garden and planting high yielding varieties while the average mulberry production per hectare has enhanced from 25,000 kg to 35,000 kgs.

Climate

Generally the Indian climatic conditions of all states are favourable to grow mulberry. The tropical temperature ranging from 24°C. But at low temperature below 13°C and above 38°C the sprouting of buds and growth of mulberry will be effected. The favourable plantation season for mulberry is early monsoon. However, if abundant irrigation facilities are available mulberry can be planted at any time. The dormant season for growth of mulberry is November to March in temperate zones and rearing could be conducted only during May to October.

In the tropical zones and rearing could be conducted only during May to October. In the tropical zones this problem does not arise and mulberry grows all through out the year under favourable conditions. Always plantation season must be selected as per the above specified recommendations; too cold or too hot seasons badly effect the growth of mulberry. The mulberry can grow very well, with in the range of 600 mm to 2,500 mm. Mulberry growth will be considered optimum in elevated areas above 600-700 MSL.

Soil

Soil matters more in the growth rate of any plant. Mulberry grows in a wide range of soils. Mulberry is a deep rooted perennial plant. Therefore, the soil with good water holding capacity and good air penetration is more suitable. Soils ranging from clayey loam to loamy soils, slightly acidic soils are ideally suitable for mulberry growth, slightly acidic soils ranging the pH from 6.2 to 6.8 reveals healthy growth of mulberry plant.

Propagation

Mulberry can be propagated in two ways: Sexual and Asexual

Sexual propagation

In mulberry the sexual propagation is through seedlings, particularly the seed propagation carries a varied population, this to utilise in selection and hybridisation. For seed germination certain prerequisites are needed to be fulfilled such as selection of quality seed, preparation of land, and the seed should be selected such that can definitely germinate. This is possible only when the seed is subjected to suitable environmental conditions, embryo of seed is alive, and healthy, in internal conditions of seed are favourable for germination.

The fresh seeds will have greater germination rate than the stored once. Seeds must be washed with fresh water until the flesh of fruit is with drawn and dried well, however minimum moisture percentage should be maintained, i.e. at least 6 %. Sowing of seeds may be by way of broadcasting or sowing in lines.

Asexual propagation

In asexual type of propagation vegetative plant parts are used. In mulberry the propagation is mainly of three types.

1. Propagation by Cutting.
2. Propagation by Grafting.
3. Propagation by Budding.

In case of mulberry, mostly through "Cuttings" only the propagation is practiced. By asexual method of propagation the inherited characteristics of parent stock can be retained. The desired characteristics can be carried to next generation. The asexual propagation in mulberry is carried out by grafting, cutting, layering.

Mulberry Cultivation

Mulberry can be cultivated in a wide range of soil. Favourable soils for mulberry growth are sandy loam, loam and clayey loam. The mulberry can be even grown on slopes of hilly areas, which are not prone to water logging. In case of slightly sloppy lands, proper drainage must be provided. No doubt the mulberry cultivation can be practiced in wide range of soils, however the leaf yield is the main factor. To increase the quality of leaf yield per hector all required favourable mulberry cultivation practices should be adopted, because the silkworm is fed by mulberry leaves only.

The soil pH, water retaining capacity of field, quality of mulberry variety, temperature, environmental conditions fertiliser/manure application, irrigation practices and soil texture etc., all these factors very much influence on quantity and quality of leaf yield per acre per year. Land should be prepared during early spring or late autumn. In Andhra Pradesh generally the most favourable season for plantation is pre-monsoon.

The proposed land should be deeply ploughed during April-May. Since the mulberry is a deep rooted perennial plant and can last even for 10-15 years. The land should be ploughed repeatedly to loosen the soil and all gravel, stones and weed should be removed making the soil fine. In irrigated land the recommended organic manure such as cattle dung compost @20 tonnes per hector or @ 10 tonnes per hectare in rainfed conditions must be applied and mixed with soil by ploughing 2-3 times.

Land levelling is also important in case of plains. However, in hilly areas generally the land will be sloppy where pit system is adopted. In hilly areas June/July are the favourable where pit system is adopted in rainfed areas. In case of pit system the pits should be filled with loose soil and organic manure.

Fertilisers Application

Fertilisers also must be used as per the recommended dosage mainly N.P.K. in case of irrigated mulberry cultivation 120 kgs N, 50 kgs of P, and 50 kgs of K per acre help for better leaf yield in Andhra Pradesh. Similarly the recommended dosage of N.P.K will be in the ratio of varied from place to place. Generally N.P.K. will be in the ratio of 2.5:1:1 for irrigated conditions and 2:1:1 of N.P.K for rain fed conditions. In hilly areas the recommended N.P.K. are 100:40:40.

Mulberry Varieties

Mulberry variety suitable to that particular area of plantation specifically recommended. For hilly areas high yielding mulberry varieties, K-2, S-54 are recommended. And for the plain areas of Andhra Pradesh, Karnataka M-5 is recommended. The varieties are recommended on the basis of favourable response in the existing environmental conditions.

Spacing

Either 3' x 3' spacing is followed in pit system or 2' x 2' system is followed in row system in rainfed conditions, under irrigated conditions, the ridges and furrows are made at a distance of 0.30 to 0.45 m. Mulberry is planted at a distance of 0.10 to 0.15 m between the plants on either side of ridges along the row.

Inter Cultivation

To eliminate the weeds and to loosen the soil for good aeration, around mulberry plant, light hoeing should be done two months after plantation. This will facilitate the plant to absorb the water freely and plant growth will not be restricted by weeds by competing with mulberry plant for food, nutrition and water.

Pruning

Pruning of mulberry trees should be done after one year of plantation. By pruning the mulberry branches, leaf yield can be increased and production of leaf can be synchronised with silkworm rearing schedules all through the seasons. Pruning schedules controls the irregular growth of mulberry branches there by save wastage of nutrition and energy. Care should be taken during pruning, bark should not get peeled off since cut wounds do not heal, which leads infections and diseases.

Under pit system of mulberry plantation pruning can be carried once in June and again in November leaving a height of 8-10 cm above the ground level in wider spacing. Further under row systems pruning should be done at a height of 3 to 8 cm above the ground level in closer spacing. Various methods of pruning are:

a) *Fist Form:* Means cutting of mulberry plant each year at one place on main stem. The part gets thick & becomes fist shaped after few years therefore, the name. Easy control of mulberry diseases & pests possible in this method.

(b) *Non-fist form:* Pruning done such that 2-3 branches allowed to grow from main stem (basal parts) & secondary branches also develop therefore, non-fist form. Resistance towards disease / pest control comparatively less in this method.

Leaf Harvesting

Along with leaf production, leaf utilisation is also important. Leaf harvesting depends upon type of rearing practice. Leaf harvested by 1) leaf picking, 2) Branch cutting / shoot harvesting. Leaf harvesting in time is very essential as mature leaf contain low nutrient value & silk worms fed upon them become disease susceptible.

Leaf Picking

In India mostly leaf picking is practiced. However, this is more labour intensive, as availability of labour is not a big problem. Presently this method is being followed by most of the rearers. In a year 5-6 harvests are possible. First leaf harvest takes place after (10) weeks of bottom pruning.

The quality of leaves harvested with petiole is better. Where moisture content/nutrient value will not be dried up. Generally the farmer picks up the leaves well in advance and feed to silk worms. Leaves are harvested according to the age of silk worms. Tender leaves are picked up, chopped and fed to chawki worms. As the stage of silkworm advances mature leaves are fed to the worms.

Branch Feeding

Silk worms fed with mulberry branches hence name – branch feeding. This type of feeding adopted after 3rd moult of silk worm larvae. Some advantages of branch feeding are:

— Cost of labour minimised.

— Quality (succulency) of leaves maintained (as leaves not detached from branches) for a longer duration.

— Bed cleaning process easy & labour saving.

— Easy to disinfect rearing equipment, if branch feeding adopted in shelf rearing / floor rearing.

— Easy to preserve & maintain quality.

— Easy to feed silk worms.

— Wastage of leaf minimised

— Hygienic conditions maintained effectively in rearing house.

— Maximum utilisation of mulberry leaves by silk worms.

Whole Shoot Harvest

This type of harvest reduces labour cost. Mulberry branches cut close to the ground level & fed to worms settled for 4th moult hence, uniformity in maturity of leaf observed. In this type, shoots harvested at an interval of 10 weeks & may extend to another 2-3 weeks depending upon availability of water and favourable climatic conditions. Around 5 harvests possible in this method.

Preservation and Transportation of Leaf

Mulberry leaf should be harvested in the early hours of the day, as the day passes the quality of leaves in terms of moisture content comes down due to active transpiration. Hence the leaves are picked up during the morning hours. Further, the leaves after harvesting should be transported in wet gunny cloth and preserved carefully, either in wet gunny cloth or leaf chamber made up of wood. Frequently water should be sprinkled on the leaf to keep the leaf fresh. However, leaf with water drops should not be fed to silk worms, which results in disease attack.

To keep the nutritive value of leaves qualitatively fresh and for healthy growth of chawki worms always tender leaves (top leaves) should be picked up. As the worms mature, matured leaves should be fed. The leaves always must be succulent, healthy cocoon production and are susceptible to disease attack. Always fresh leaves are recommended to feed the worms. Longer the preservation, lesser the moisture and the nutrition, hence temperature, humidity and moisture contents in the storage place should be 20° C–90% of relative humidity and cool, clean with frequent watering.

Silkworm Rearing

Silkworm is domesticated over thousands of years and the sole food for the growth of silkworm is mulberry leaf.

Climatic requirements

Under ideal conditions silkworm completes cocoon formation in 24-28 days from the day of hatching. However, there is variation in number of days due to fluctuation in temperature and humidity. As the temperature increases the development of larvae activates. The following required temperature/humidity/ spacing should be provided:

S. No.	*Stage*	*Temperature °C*	*Humidity %*	*Spacing (for 100 dfl's) in Sq. ft*
1	1st Instar	26-28	85-90	4-14
2	2nd Instar	26-28	85-90	15-45
3	3rd Instar	25-26	80-85	46-90
4	4th Instar	24-25	70-75	91-100
5	5th Instar	23-24	70	181-360

The larval development will be maximum during the 5th Instar after 4th moult. Generally the moulting period delays due to fluctuation in climatic conditions. Temperature plays an important role in the development of larva. Larva under goes four moults in case of tetramoulters with in a span of 26-28 days to spin the cocoon.

Rearing Shed

Rearing shed must be constructed with East/West direction, preferably with the thatched roofing, mud walls or any roofing must not reflect heat on the biological body growth of silkworm. Keeping good number of windows, proper ventilation, light, using country tiles for roofing will serve the purpose of maintaining required atmospheric conditions in the rearing house with reduced cost. The rearing house should have a good length of height and breadth.

Quality of Leaf

Quality of mulberry leaf in addition to proper maintenance of temperature/humidity plays significant role in the development of healthy silkworm. Leaf quality plays an important role in the production of quality cocoons. The young age worms are fed with tender, succulent leaves which contain sugar, less amount of fibre, starch but, high moisture and protein that are suitable for chawki worms. Hence, the top tender leaves of mulberry plant are used to feed the young age worms. The late age worms are fed with matured leafs available at the bottom of the mulberry trees.

The leaf during its feeding time should not contain dust particles or water droplets on the surface of leaf. Disease attacked, wrinkled, weak leaves, poor quality leaves should not be fed to the worms. Feeding bottom matured leaves to chawki worms, tender leaves to mature worms should be avoided.

Leaf Preservation

Harvested leaf should be transported in wet gunny bags or baskets made up of bamboo. Such leaf should be preserved in a separate room or in a corner of rearing room or in specially designed leaf preservation chamber made up of wood with sufficient number of ventilators. The leaf chamber should contain a good number of ventilators covered with wet gunny cloth. This will help in maintaining leaf quality.

Other wise leaf should be scattered on the floor in a single layer so that the water droplets on surface of leaf will be evaporated. Leaf should never be preserved on heaps. This will increase the humidity. Leaf scattered on floor should be frequently sprinkled with water and tilted in summer.

Leaf Requirement

Silkworm attains nearly 10,000 times of weight starting from hatching to spinning stage, therefore, feeding quality leaf plays an important role in the development of silkworm.

Leaf Requirement (about 400 eggs per DFL's)

S. No.	*Silkworm Stage*	*Quantity of mulberry leaf Required (approx.) (Kgs)*
1	1st Instar	2-4
2	2nd Instar	4-8
3	3rd Instar	30-40
4	4th Instar	80-90
5	5th Instar	600-650

The requirement of leaf will differ from race to race. The consumption of leaf also varies with season. During summer the quantity of leaf required will be a little higher since, drying of leaf will be faster hence additional feeding may be required.

Rearing equipments

The following equipments are required for silkworm rearing.

1. Mesh
2) Formalin
3) Sprayer
4) Mats
5) Leaf preservation chamber
6) Chopping board
7) Chopping knife
8) Chop sticks
9) Feathers
10) Ant wells
11) Foam pads
12) Paraffin Paper
13) Hygrometer
14) Thermometer
15) Bed cleaning net
16) Wash basin
17) Plastic buckets/Mugs
18) Rearing stands
19) Rearing trays
20) Feeding stand
21) Mountages

Disinfection

Disinfection is a process of destructing the disease casual organisms. There is a possibility of carrying the pathogens or germs through the rearing equipment if there is any disease attack in the previous crop. Therefore, disinfection is necessary to protect the crop from disease attack. Before taking up rearing all the rearing equipments including the rearing house should be disinfected thoroughly.

There are various methods of disinfection like physical and chemical. Physical disinfections are Sun drying, Steaming. In chemical disinfection, all the rearing equipment as well as the rearing house should be thoroughly disinfected with 2% formolin and dried. Ultimate production of silkworm seed is to produce cocoons to get silk yarn. Advance planning, quality production, timely supply of seed are the critical factors in silkworm seed production.

The silkworm seed is produced in "Grainages". Grainages play a vital role not only in production of seed but also in the entire seed organisation. During the Grainage operation one must know the seed suitability for a particular area before releasing commercially. A particular race is suitable to a particular area.

Hybrid Varieties

CSR & TI have evolved certain high yielding varieties and recommended to different regions suitable for different seasons.

Hybrids	*Suitable Season*	*Suitable Region*
F2D1 x NB18	Winter,Spring	Andhra Pradesh, West Bengal, Orissa, Madhya Pradesh

MY1 x NB 18	Spring/Autumn	West Bengal, Assam, Bihar
N x (NB 18 x P5)	Autumn,Summer/ Autumn	West Bengal, Orissa, Madhya Pradesh
SH6 x KA	Spring/Autumn/ Winter	West Bengal, Orissa, Madhya Pradesh
CA2 x NB4D2	Spring Spring/Autumn/ Early Winter	West Bengal, Orissa, Madhya Pradesh J & K / Uttar Pradesh
PM x NB 18	Summer	Orissa, Madhya Pradesh
NB 18 x P5	Winter	Orissa, Madhya Pradesh
YS 3 x SF19	Spring	J & K / Uttar Pradesh
SH6 x NB4D2	Spring	J & K / Uttar Pradesh
PAM101 x NB4D2	Autumn/Early Winter	J & K /Uttar Pradesh
CC1 x NB4D2	Autumn/Early Winter	J & K / Uttar Pradesh
PAM111 x SF19	Autumn/Early Winter	J & K / Uttar Pradesh
RD1 x NB18	Summer/Early Winter	Uttar Pradesh

Transportation of Eggs

The required number of DFL's should be collected from either Govt. Grainage or licenced Pvt. Grainage. The transportation of eggs plays an important role in the development of embryo and successful crop harvest. Therefore, the DFL's are collected safely in a wet hand bag in the early morning or in the late evening. Care must be taken to not to cause any damage to the inside embryo.

In case of loose eggs, the transportation box containing loose eggs should be covered with wet cloth, good aeriation must be provided to the eggs. Care should be taken while transporting the DFL's and should be kept under incubation at 25°C temperature and 80% humidity.

Egg cards should be spread in the rearing trays which should be kept in cooler places only.

Incubation

The eggs should be kept in cooler places at 25°C temperature and 80% humidity, lesser the humidity poorer the hatching percentage or more the temperature weak the larva and poorer the hatch percentage. For uniform hatching all the egg cards or loose eggs should be kept in dark and cooler atmosphere. On the day of pin head or blue egg stage all the eggs are covered with a black sheet or kept in black box and known as black boxing.

On the day of hatching all the eggs are suddenly exposed to bright light in the early morning at around 8 A.M. so that 95% hatching can be achieved. It should be ensured that hatching should take place uniformly. With proper maintenance of temperature and humidity development of larvae will be uniform.

Brushing

Brushing is transferring of newly hatched larvae into rearing trays. The newly hatched larvae after one hour of hatching get ready to feed on mulberry leaf. They are fed with finely chopped tender mulberry leaf. They are fed with finely chopped tender mulberry leaves. Chawki worms are fed with tender leaves with high moisture content. The brushing of newly hatched larvae can be done in various methods.

Newly hatched larvae can be covered with a net and chopped mulberry leaf can be sprinkled over the net. The larvae slowly crawls on to the net and start to feed on the mulberry leaf. Then transfer into rearing tray by gently tapping the net, this method is more convenient in case of the loose eggs. In case of egg cards, the egg cards are placed in the rearing trays and chopped mulberry leaf is sprinkled over the newly hatched larvae, the larvae crawls on to the mulberry leaf later on the cards are removed. Care should be taken not to touch the newly hatched larvae with hands while providing proper spacing, instead chop sticks are used to spread the worms in the rearing tray.

Leaf Feeding

The leaf is finely chopped as per the recommended size and sprinkled on the larvae and after half an hour the newly hatched larvae slowly crawl on to mulberry leaf. Chopped leaf should be preserved from drying and loss of moisture content. For this purpose it should be collected in a basin covered with paraffin paper to protect from loss of moisture content. First feeding to the newly hatched larvae should be given at 9 A.M. in the morning.

During the 1st instar depending upon the silkworm strain and other factors (environmental conditions and leaf quality) the required mulberry leaf per 100 DFL will be around 2-2.5 kgs. The 1st instar stay for 3-31/2 days and under go moult. The moulting period last for about a day. This will depend upon the environmental conditions. Gap should be maintained for feeding the worms before under going moult and immediately after moult. Generally two hours gap is given. Also over feeding should be avoided, especially before under going moult which will result in wastage of leaf.

Spacing

Larval growth will be very fast in the silk worms during the early stages. Too much crowd in the rearing tray results in increase in humidity, heat, fermentation of litter which results in un-hygienic conditions, wastage of leaf and under development of silkworm. The number of trays and space required for each instar will increase.

Stage of silkworm	*No.of trays required for 100 DFL's (Diameter of tray 3½')*		*Space required (Bi and Multi x Bi hybrids)*	
	Beginning	End of stage	Begin	End
1st instar	2	2	4 to 15 Sq.ft	
2nd instar	2	5-6	15 to 45 Sq.ft	
3rd instar	5-6	10-12	46 to 90 Sq.ft	
4th instar	10-12	20	91 to 180 Sq.ft.	
5th instar	20	40	181 to 360 Sq.ft	

Spacing should be given simultaneously at the time of bed cleaning or feeding and care should be taken not to disturb the worms very frequently.

Bed Cleaning

Silkworm after moulting leaves the exuviae. All the leaf fed to silkworm may not be consumed. Dried leaves, rejected leaf in the bed, silk worms excreta, dead worms, diseased larvae all will increase the humidity, fermentation, temperature in the bed. If proper bed cleaning is not done in time it leads to various complications viz. Ill health of larvae, disinterest of the larvae to fed due to unhygienic conditions, ultimately worms becomes weak. Bed cleaning in silkworm rearing tray is done by various methods like using of paddy husk, straw, and bed cleaning net.

During 1st instar bed clearing should be done once during pre-moulting, during 2nd instar – twice, once after moult and before next moult. During 3rd instar thrice i.e. after moult, before next moult and once in the middle. During 4th and 5th stage once in a day in case of shelf rearing. However, in case of floor rearing or shoot rearing once in each instar.

Feeding

Not only the nutritive value but also the number of feeds in each instar plays a major role in the cocoon built. Three to four feeds are given to the silkworm. In a day the last feeding i.e., during the night should be a little more since the duration for the next feeding will be longer. During moulting period no feeding should be given. Before settling to moult and immediately after moult, first feeding should not be heavy.

Feeding tender leaves to young age worms is essential. As the larval growth advances the mature leaf can be fed. Maximum leaf is consumed during 5th instar only. During this stage only the maximum growth of silk gland can be noticed. It is estimated that 50% of the total weight will be increased in the 5th instar itself. The feeding schedule is: -

S.No.	1st feeding	2nd feeding	3rd feeding	4th feeding	5th feeding
1	6 A.M.-6.30 A.M	11-11.30 AM	3-3.30 PM	7-7.30 PM	
2	7 AM – 7.30 AM	11-11.30 AM	2-2.30 PM	5-5.30 PM	
3	6 AM – 6.30 AM	10-10.30 AM	2-2.30 PM	6-6.30 PM	10-10.30 PM

More number of feedings are given during summer since moisture content in the leaf will not be sufficient during summer and leaves will dry very easily.

Types of Silkworm Rearing

Rearing of silkworm is done in various methods in different areas. Mainly the rearing of silkworm is of two types:

Chawki rearing: Rearing of young age silk worms is called Chawki rearing. Here worms are reared upto third moult and distributed to the rearers for late age rearing. Chawki rearing in mass at each CRC will not only control the attack of diseases but also facilitate the rearing on most scientific lines and cost wise it is most economical. To maintain coolers, heaters, exhaust fans, etc., by individual rearers is a costly affair, however, all these facilities can be maintained in Chawki Rearing Centres.

Late age rearing: Late age rearing after third moult does not require high temperature and humidity compared to chawki rearing. Late age rearing is a little easier process than chawki rearing. During late age the quantity of mulberry leaf required is more than 90% of total larval period. During 5th stage particularly the larvae eat voraciously worms feel maximum appetite larvae loses water from its body hence, less temperature, low humidity, good ventilation is required.

Mounting

This is the last stage of rearing operation. At this stage silkworm becomes matured completely and ready to spin into cocoon. These silk worms after 5 days of 4th moult becomes yellowish and translucent, stops feeding on mulberry leaves, starts to search for a corner and try to crawl on the feeding trays further liquid like substance oozes out of the mouth from the spinneret. These movements clearly indicate to transfer the mature larvae into the mountages.

Transferring of matured silkworm on to the mountage or cocoon frames is called "mounting". This is an important and skilled operation. Any deviation in identification of maturity of worms adversely effect on cocooning. During mounting though, selection of ripen worms is not a difficult task, but, due care is needed. Most of rearers pick up ripen worms with hands to transfer them on to mountage.

A few rearers use rearing nets, in some parts, rearers use mulberry branches to separate the matured worms. As the matured worms stop feeding they crawl on to the twigs searching for a support to spin the cocoon. Another method called frame method is also used, fresh mulberry leaf is applied with a frame, matured worms crawl above the frame which are picked up and tranferred on to the mountages. While mounting care should be taken to avoid over crowding. Proper spacing should be given to avoid formation of double cocoons strained cocoons/ urinated cocoons.

For the purpose the density of worms in the mountage should be limited to 40-50 worms per sq.ft. to avoid formation of double cocoon, stained cocoon. However, the number of worms in each mountage depends upon the size of mountage and type of silkworm race. The matured worms picked up in any of the methods mentioned above are transferred to the mountages. In an area of 1 mt 800-900 worms can be mounted to avoid too much density in the mountage.

Mountage should be kept in inclined position so that, excreta of worms fall on ground. In certain strains of silk worms around 250-300 worms can be conveniently mounted in a standard mountage of 90 x 60 cm size. However the size of mountage varies from place to place. In case of bamboo made chandrike that has a mat on the back ground of size 6′ x 4′, it easily carries above 1000 worms. During mounting care should be taken to use the right type of chandrike in convenient size and shape. There are different types of chandrike that are used for the purpose of mounting.

1. Plastic mountage
2. Bamboo made chandrike
3. Straw mountage
4. Bottle brush
5. Revolving mountage, etc.

Cocooning

The matured silkworm spins the outer protective covering called cocoon and remains in dormant stage inside as pupa. After mounting the ripen worms in chandrike the larvae sticks on to the mountage by oozing out the silk fluid which will harden immediately after coming in contact with air and sticks itself to the mountage. It starts to ooz out the silk by continuous movement of its head in a very specific manner to form the silk filament in the shape of arithmetic figure '8'.

Larvae moves its head continuously about 70-80 times per minute till the compact shell called cocoon is formed and detaches itself from the last layer of silk of then transforming into pupa. The Cocoon will have three layers.

a) Outer floss

b) Middle compact layers

c) Inner pelade

The outer floss layer which will be 8-10% of weight of cocoon though formed of a continuous filament, is not reelable. The floss percentage varies from race to race. The middle order compact layer only is reelable and is a continuous filament easily reelable and the inner pelade layer is not reeled. Silkworm completes the spinning in 2-4 days depending upon the silkworm strain and climatic conditions provided.

Harvesting

Harvesting of cocoons is done on the fifth day of spinning. Whereas seed cocoons should be harvested on eighth day or ninth day of spinning depending upon atmospheric temperature. Harvesting should not be done immediately after pupation. Further, harvesting should be done before the moth emerges out. Too many days delay in harvesting will result in formation of pierced cocoons due to emergence of adult moth or uzi maggots. Cocoons are harvested generally with hand. In advanced temperate countries like Japan simple devices are used to harvest the cocoons from rotatory mountage.

Indian silk suffers from variation in the denier which breaks during winding and weaving and thus is not preferred as a warp in the powerloon, while the chinese silk having betters strength, colour and shine is used without twist for higher production efficiency. If Indian silk can be better with respective denier uniformity and strength, it can be used in powerloon as it has a better luster and dye affinity. At present, multi end machine are not installed in any of the big factories but are used by the small reelers who can produce only a small quantity of multi-bivoltine silk.

What is needed is private sector investment or producing large quantities of row silk (C-2A grade) to feed the powerloon sector. The bivoltine hybrids which promised quality, silk as well as increased production have not been explotted in this decade. Therefore it is necessary to switch over to sustainable highly productive bivoltine silkworm breeds which can give 2A-4A grade international quality silk. The rearers will get the price for their product depending upon the quality of cocoons they produce.

The main considerable commercial characters of cocoons for price fixation are:

Colour: The colour of cocoon depends upon the presence of coloured pigments in the sericin layer. Properly stifled cocoons will be brighter and dullness indicate the improper stifling.

Compactness: Compactness of cocoon is an important character that indicates the quantity of silk present in the cocoon. Compact, slightly resilient firmness of cocoon when felt by light press indicates the good commercial quality of cocoon.

Cocoon weight: The weight of cocoon is one of the important commercial character considered in price fixation. The green cocoon weight will be decreased day by day until moth emerges out.

Denier: Thickness of silk filament is denoted by denier. The denier size of outer most layer of cocoon i.e. floss layer will be higher than the inner most pelade layer.

Grain or wrinkle: A deflossed cocoon should not show grain or wrinkles. Which are not easily reelable. Cocoons with finer granulations are easy to reel.

Length of silk bave:

— Indian multi voltine races contain 300-400 mts

— Indian multi voltine hybrids 400-550 mts

— Newly evolved hybrids 600-800 mts

— Uni/Bi-voltine hybrids 1000-1500 mts

Reelability: Reelability is the ratio of cocoons reeled without break and the total number of cocoons utilised. Frequent breakage results in wastage of raw material and wastage of time, hence, skill labour, quality cocoons are essential.

Shell atio: The quantity of silk can produced from each cocoon depends upon the weight of shell. Therefore, necessarily it is important to calculate the shell ratio. The cocoon weight includes the weight of shell and weight of pupa inside.

S.R.%= Wt.of cocoon shell/Wt. of cocoon x 100

Shape: There are different shapes of cocoons; round, oval, spherical and pointed ends, and spindle shape. Spherical egg shaped and moderately constricted cocoons are easily reelable, where as deeply constricted in the middle and cocoons with pointed ends are commercially not suitable for steady reeling.

Keeping in view of the above important characters for commercial purpose the defective cocoons such as undersized cocoons, malformed cocoons should be avoided due to the less content of silk. Immature cocoons stained cocoons and cocoons having mould should be avoided, as it is results in poor reelability and wastage due to the attack of fungus.

Reeling

The process of drawing silk fibre from the cocoon is called 'reeling'. The cocoons are cooked in hot water and the silk fibre is unwound from the cocoons. The silk consists of two proteins, the inner core of fibroin and an outer cover of gum sericin. During reeling, the cocoons are processed in hot water at 95-97°C for 10-15 minutes. This process is called cooking. This cooking will enable the sericin portion to get softened and make unwinding easy without breaks.

The cocoons after cooking are reeled in hot water in different types of machines. In India, 61% of the silk amounting to 1,320 tonnes is reeled on the country-type charka (spinning-wheel) numbering 6,656. The silk produced with the country charka is of very poor quality, as the thread is not uniform, as it carries many slubs etc. the improved cottage-type basins have been introduced recently into India. Provision for button-holes and a proper croisure system to maintain the thickness of the fibre, and to control the defects of neatness in the cottage basin have facilitated the production of better-quality silk. As much as 4,000 cottage basins in the country produce 806 tonnes of silk.

Large-scale basins organised scientifically are arranged in filatures for the extraction of superior-quality fibre. The silk produced by the filatures is superior because of the low level of defects of cleanliness and uniformity in the thickness of the fibre. Only 8% of the total production of silk in India is contributed by filatures. The silk produced from multivoltine races of silk worms is poor in quality and is known to have greater defects, such as lousiness, and defects in neatness and cleanliness and is of very poor quality in light of the international D grade.

The silk produced by the bivoltine races of silkworms possesses superior neatness and cleanliness, is without lousiness and has high tensile strength and stands to the international A grade. The new technology of handling silkworms in the country has shown that bivoltine silkworms, producing the international quality of silk, can be produced throughout the year in the Karnataka state in India. It has been shown that, on an average, 30 to 35 kg of cocoons, yielding 3 to 4 kg of high-grade silk can be produced by rearing 40,000 eggs of bivoltines, as compared with 25 kg of cocoons and 1.4 kg of low-quantity fibre from multivoltines. Factors affecting the quality of silk are:

Water: The water used in silk reeling plays an important role. It is estimated that to produce one kg. Of silk from charka, the volume of water required is about 100 liters. If the quality of water is poor, the impurities suspended in the water will adhere to the silk thereby the colour, and lustre of silk would be influenced. The water used for cooking cocoons should be colour less, odour less and transparent. The pH of water should be between 6.8 to 8.4.

Cocoon drying: The cocoons contain live pupae which will emerge out with in 10-12 days of cocooning hence, cocoons ae stifled. Once the moth emerge out, the shell becomes of no use to reel since the adult emerges out by piercing the cocoon shell. The cocoons contain live pupae, which will emerge out with in 10-12 days of cocooning hence, cocoons are stifled. Once the month emergesout, the shell becomes of no use to reel since the adult emerges out by piercing the cocoon shell. The primary object of drying cocoons is to kill the pupae and to reduce the water content of fresh cocoons.

Methods of Stifling

Sun drying

The freshly harvested cocoons are directly exposed to sunlight. In this method the harvested cocoons are spread in one layer and directly exposed to bright sun-light for several days from morning to evening until the pupae inside is killed and cocoon is found properly dried. This can be identified by shaking the cocoon which will make rattling sound. Some advantages are:

1. This is cheapest method of stifling
2. Easily available method of cocoons drying to rearers.
3. These cocoons can be reeled immediately.

Some disadvantages are:

1. Continuous exposure to hot sun-light, damages the silk quality due to the action of ultra violet rays.
2. Some times continuous sun-light may not be possible, particularly during rainy season. Partial drying of cocoons will spoil the quality of silk.
3. Wastage of silk during reeling will be more.

Steam stifling

Methods

— Steam stifling is the very common and easy method.

— The fresh cocoons are directly exposed to hot wet steam.

— Basket steaming

— Barel steaming

— Chamber steaming

Advantages are:

i) conveniently used to stifle large quantities of cocoons

ii) uniform stifling of cocoons

Disadvantages are:

i) steam stifled cocoons cannot be reeled immediately, other wise the sericin will wet and fibroin comes is lumps while reeling

ii) in steam stifling pupa is killed inside and does not dry it.

Hot air conditioning

In the modified hot air conditioning chamber there are

five compartments. Each chamber is different ranges of temperature.

— The first chamber is filled with cocoons are supplied of with hot air at 50°C for about 3hrs.

— The second chamber by sliding mechanism by pulling out where 70°C for about 3hrs.

— The third chamber cocoons are dropped into 75-80°C.

— The forth chamber is cocoons dried by hot air at 80-90°C temperature

— The fifth chambers are gradually exposed to hot air in the 12 hours of time and continuous process.

Advantages are:

i) Any amount of cocoons can be easily stifled.

ii) Time saving, labour saving.

Some disadvantages are:

i) Cocoons of uniform size are quite suitable for this process, but Indian multi-voltine cocoons with different sizes and different qualities need different degrees of temperature for stifling.

ii) Costly and skilled labours are required.

Storage of Cocoons

While storing the cocoons certain precautions should be taken into consideration other wise improper storage of cocoons leads to loss due to occurance of damage to the cocoons. Preventive measures should be taken:-

1. Avoid storage of stained cocoons
2. Storage room should have good ventilation.
3. Cocoon storage place should be clean and dry. No waste cocoons should be stored in the premises.

4. Always ensure proper drainage of cocoons before storage.
5. Relative humidity in the storage room should not exceed 70%.
6. Before storage of cocoons, the storage place should be thoroughly disinfected.
7. In case of long storage of cocoons, frequent turnings should be given.

Sorting

Sorting of cocoons is already done at once, once again it should be checked prior to reeling.

Deflossing

Floss is the outer most layer of cocoon which is only a protective covering and is not reelable. Prior to reeling floss has to be removed by any of the methods to avoid obstruction during reeling.

Ridding

After deflossing the cocoons, in the process of ridding, cocoons are separated according to their sizes by simple sieves into big, small and medium size cocoons and collected in separate containers. This will facilitate the reeling operations easy.

Reeling

Reeling is a process of unwinding of silk filament from the cocoon. Reeling process is an important activity. Depending upon the required thickness (denier) of silk thread filaments from number of cocoons are combined together and reeled. An efficient reeler will maintain the fixed number of cocoons per end to produce uniform denier silk. Reeling is carried out by distinctive methods.

Types

Charka

It is about 50% of total raw silk production is contributed by charka. Reeling machine is traditionally home built by using wooden material with the assistance of black smith and corpenter hence,

1. Investment on reeling machinery is very-very less.
2. Even the inferior quality cocoons can be also reeled by charka.
3. No much technology is involved in this system of reeling.
4. Only one reeler can reel.
5. Brushing and cooking are done in the same basin.
6. The cost of production is very low compared to other reeling machines.
7. Involvement of labour is very less only two persons are sufficient to maintain.
8. Reeler can establish a charka unit at the back yord of his own residence.
9. Production per day would be around 1 kg from each charka.
10. Marketing of charka reeled silk is very easy and it will be always at his door steps.

In charka, there are three distinct parts.

1. Mud-Platform
2. Distributor
3. Reel.

Some advantages are:

1. The reel should be always smooth.

2. As the cooking and brushing, reeling are done in the same water. The water should be changed frequently.
3. As the cooking and brushing, reeling are done in the same water.
4. After cooking the cocoons thoroughly, when reeling ends are found, the temperature of water in the reeling basin should be lowered to 40°C by adding cold water.

Some disadvantages are:

1. The silk produced by charka reeling is inferior quality.
2. As the reeling is done in water at very high temperature which is harmful to the reeler.
3. Frequently, water is to be changed and heated again, this will increase the fuel cost.

Cottage basin

In cottage basin, the cocoons are cooked separately and re-reeling is done separately to increase the quality of silk. In cottage basin cooking is done separately and the reeling basins are fixed to a reel bench. There will be 6-10 ends in each basin. Arrangement to supply hot water to each basin and to drain out the dirt water accumulated due to floss and sericin is made. Some advantage are:

1. Quality of silk produced is superior to charka silk.
2. Water in the reeling basin will be fresh. This will help to maintain the quality of silk.
3. Production per day can be increased, as it is power operated.

Filature

Filature or multi-end reeling machine works on the principle of slow motion reeling and thread production on small reels at a large number of ends per basin. Some advantages are:

1. Superior quality silk can be produced
2. Wastage is reduced
3. High productivity
4. Less labour
5. Arrangements are made to control the speed of the reels and tension of the threads.
6. Raw silk per day will be around 800 gms of superior quality.

Automatic reeling machine

Bi-voltine cocoons are best suited to automatic reeling machines. This machinery requires superior quality cocoons and uniform size of cocoons with less floss. Generally the multi-voltine cocoons are not fit for automatic reeling. This is particularly due to inferior quality. The silk produced from automatic reeling machine will be superior.

Silk Knit Fabric

Silk knit fabrics could contribute 5-10% to the total export of silk materials, currently worth Rs.1500 crores from the country. Silk knit T-shirts, sweaters and children wears have tremendous scope and huge export demand in Europe, the US, Japan and China is a major contributor. India had the advantage of the availability of moisturous and lustrous looking raw materials. The Bapuji institute of Engineering and Technology had developed silk knitted fabrics after conducting a survey with knitwear exporters and manufacturers across the country, including Tirupur.

The institute was able to develop 40 different type structures including blended once, with its new technology, with the assistance of the Swiss Agency for Development and Co-operations. The Indian silk woven materials were sold for $13 per 100g, knit wear could be sold as low as $8. Thus it had value-addition 10 times more than cotton. Bapuji Institute of Engineering and Technology in Karnataka was also exploring the possibility of establishing some depots near Tirupur, which was away from the silk sector, but had other advantages such as facilities for dyeing and processing.

Such 12-15 firms established in the area had obtained samples and sent to various countries. It can be seen that in future the demand for Indian wears would also increase. About 15,000 million tonnes raw silks were produced in the country, besides 8,000 million tonnes imports. 75% of Indian production were used domestically and only 25% for export purposes. Silk knit was not necessarily a winter or summer fabric and therefore could be used round the year depending on its weight and knitting construction. The Institutes can definitely beat China, as it believes in bulk supply rather than small orders.

Different Consumption Trends

Karnataka accounts for 25% of the country's silk export. This is valued at Rs.400 crore currently but would hopefully go upto Rs.1, 500 crore by 2010. The state of Andhra Pradesh has planned to double the area under mulberry cultivation from the existing 1.10 lakh acres to 2.48 lakh acres by 2020, in line with the Vision – 2020 plan. The Indian govt. has launched an ambitious programme called – operation silk wave to help in the development of superior bivotine silk in the country.

It is expected that by 2007 production of bivoltine silk would touch 10,500 tonnes from the 300 tonnes being produced currently. Simply put this means reducing the demand supply gap in case of raw silk from the existing 48 per cent to less than 20 per cent in 2007. The focus of Operation silk wave is to help raise the production of bivoltine silk in the traditional strongholds of mulberry silk, namely—Karnataka, Andhra Pradesh and Tamil Nadu.

All India raw silk productivity has doubled in about five years time. In 1995-96, it was pegged at a lot 32% hectare but recorded a near 50% jump to touch 46kg in the very next year. It has since then recorded steady upward movement and was placed at 61kg per hectare in 1999-00. The Karnataka State was planning to raise the area under mulberry cultivation (notably of the new improved varieties like V-1 and S-36) to 50% of the total area under bulberry cultivation.

Incidentally the area under mulberry cultivation rose during 1990-91 to touch 1.49 lakh hectares against 1.14 lakh hectares in 1980-81. Coinciding with the rise in area under mulberry Karnataka also witnessed a near 50% rise in cocoon production, which touched 59,033 tonnes during 1990-91 as against 38,066 tonnes in 1980-81. Production can be increased with irrigation facilities. From the existing 300 hectares under drip irrigation the state plans to rise this level to 2,500 hectares by 2010.

The state also plans to set up 5,000 independent rearing houses against 100 at present. Production of bivoltine silk is projected to go up from 212 tonnes presently to around 10,000 tonnes in ten years time. The state, which has also enacted the Reshme Sampathu

Yojane is planning to invest a sum of Rs. 400 crore over the next five to six years and plans to seek financial assistance from various agencies including the World Bank and the Govt. of India.

The more ambitious plan in Andhra Pradesh's scheme of plans happens to be in increasing production of bivoltine silk production from the existing 30 tonnes to about 94,000 tonnes by 2020. Multivoltine silk production is also expected to rise to 1.44 lakh tonnes from the existing 30,000 tonnes. In the other major producing state, Tamil Nadu, the area under mulberry cultivation was 12,900 hectares during the 8th five-year Plan, which is expected to touch 12,550 hectares by 2002.

By 2007 the area would witness a rise by about 66% touching 20,550 hectares. Production of raw silk, which declined from 774 tonnes during the 8th plan Period to 672 tonnes. It is estimated that raw silk production would go upto 1,800 tonnes by 2007 up from 800 tonnes in 2002. Production of bivoltine silk is likely to witness a five-fold increase from 55 tonnes (in 2002) to 300 tonnes in 2007.

Silk Export

Japan, China and India are the leading countries in the production of silk and other silk materials but now a day there is a greater competion between China and India in case of silk production. By knowing the nerves of foreign marketing and factions by India there is increase in exporting of silk during first six months of 2000-01. By comparing last years exports of silk items, it increases from 15 01 to 16 1 and earning $17.43 crore. That means it is increased by 30.5%. There is demand for Indian silk items from America, Japan, Spain, Germany, Italy and east Europe.

Indian requirement of silk is 22,000 tonnes while its production is 15,000 tonnes. Second most important thing is to improve silk quality, which includes soft, strong and continuous thread. Though there is an increase in plantation of Mulberry by 5 times and many schemes implemented successfully for silk production there is shortage of silk. Due to shortage of raw silk there is reduction in spun silk and Kashmiri Carpets. In India 40% production of Mulberry Silk is from Karnataka. While Tasar Silk is from Madhya Pradesh and Orissa and Muga Silk is from eastern parts of India.

There is increasing demand for silk cloths from foreign markets. If we concentrate on this demands there is scope to increase exporting of silk items up to 1500 crores. Following table shows exporting of silk from India.

Year	*Export (Rs. in crores)*
1994-95	926.96
1995-96	845.16
1996-97	876.46
1997-98	905.42
1998-99	1053.86
1999-00	1100.00

A scheme "Operation silk wave" is developed by the Central govt. for the development of quality "Biovoltaine silk". Under this production of silk will be increased from 300 to 10,500 tonnes during the period of 7 years and there is steps taken to reduce the shortage of silk. This scheme will be implemented in Karnataka, Andhra Pradesh and Tamil Nadu. Karnataka govt. has also implemented "Reshme Sampathu" scheme along with this scheme.

Silkwave

Statistics of exports issued by the Central Silk Board (CSB) indicate a robust growth during the first half of the current fiscal. In dollar terms this worked out to $174.82 million against $150.11 million for the same period in FY '98-'99. Value added products like shirts and scarfs rose by a huge 213.2%. The rise in exports could be attributed to a good understanding of the international fashion market. However, the biggest worry continues to be the inadequate availability of quality raw silk.

Current demand is placed around 21,000 tonnes, domestic production being placed at about 14,000 tonnes, leaving a deficit of 7,000 tonnes. Under the operation silkwave production of bi-voltine silk is expected to touch 10,500 tonnes by '07 with lead being taken in the three southern states of Karnataka, Andhra Pradesh and Tamil Nadu. Karnataka, which accounts for over 60% of the silk production in the country, plans to raise the production of biovoltine silk from the existing 212 tonnes to over 7,000 tonnes over an 8-10 year timeframe.

The Karnataka government has also indicate that the existing state Silkwormseed, cocoon and silk yarn (regulation of production, supply, distribution and sale) Act of 1959, is expected to bring more investments into the silk industry. As a matter of recognition of India's growing stature in international market, the country has been chosen to host the 23rd Congress of the International Silk Association.

New Methods in Sericulture

Mulberry cultivation is done in the places like Solapur, Nagar, Latur, Parbhani, Aurangabad and Vardha. New intercropping system helps to reduce the water requirement and also the expenditure required on it. To increase the area under mulberry cultivation and to increase production of silk the Mulberry silk training centre, silk fibre training centre, silk fibre spinning centre and Research centre have been developed wherein information with respect of sericulture will be provided.

In future centres will be developed at Pune and Aurangabad, which will help to take the cocoons after its production. This year silk Mandal has developed varieties like CAIR Dubar, Mulberry's V-1, S-36, S-54, which will help the farmers to gain more profits. It was also told that government will give seed of mulberry at the rate of Rs 200/ acre. Mulberry crop gives production four times during the year time. It also does not have any adverse effect on the texture of soil. For an acre of land farmers can get Rs. 40,000 while the expenditure is about 20% and hence is beneficial to farmer.

Bibliography

Balanay, P. G., "Performance evaluation of newly F1 silkworm hybridsunder laboratory condition", Science Research Specialist II/PTRI TC, Kalingagan, Villanueva, MisamisOriental, Philippines.

Chatterjee S.N, Mohandas T.P, Taraphdar T., "Molecular characterization of the gene pool of Exorista sorbillans (Diptera:Tachinidae), a parasitoid of silkworm Bombyx mori, in India", *European Journal of Entomology*, 2003.

Coldrey, Jennifer, "Silkworm Story", *Film*, London, UK: Oxford Scientific Films, 1983.

Cooper, Elizabeth K., *Silkworms and Science: The Story of Silk*, New York: Harcourt, Brace and World, 1961.

Dallinger, Jane, "Silkworms", *Photographs by Satoshi Kwibaysha*, Minneapolis, MN: Lerner Publications, Co., 1981.

FAO, "Sericulture manual 2—Silkworm rearing", *FAO Agric. Serv. Bull.*, 1978.

_______________, "Sericulture manual 3—Silk reeling", *FAO Agric. Serv. Bull.* 15/3, 1979.

Johnson, Sylvia, *Silkworms*, Minneapolis, MN: Lerner Publications Co., 1982.

Krishnaswamy S., "New Technology of Silkworm Rearing", *Bulletin- 2*, Publisher: CSR &TI, Mysore, India, 1978.

Liguan, R.M. and Sanchez, Z.L., "Double Cross Silkworm hybrids for higher silk productivity", Philippine Textile Research Institute

Puttaraju H.P, Madhu M. "Presence of Wolbachia endosymbiont in different silkworms and races and their uzifly parasites", *Journal of Invertebrate Pathology,* 2002.

______________, Prakash B.M., "Effects of Wolbachia in the uzifly, Exorista sorbillans, a parasitoid of the silkworm", Bombyx mori. 7pp. *Journal of Insect Science,* 2005.

Sanchez, Z.L, "Upgrading and optimization of silk hand reelingTechnology", Senior Science Research Specialist/ PTRI TC , Kalingagan Villanueva, Misamis Oriental Philippines.

Wu Pang-chuan and Chen Da-chuang, "Silkworm Rearing", *FAO Agricultural Services Bulletin* 73/2, MF 33-809, book, 83 pages, 1988.

Index